Genetic Algorithms

Edited by Sebastián Ventura,
José María Luna and José María Moyano

Published in London, United Kingdom

Genetic Algorithms
http://dx.doi.org/10.5772/intechopen.94664
Edited by Sebastián Ventura, José María Luna and José María Moyano

Contributors
S. Tamilselvi, PhamThi Ly, Bui Quoc Khanh, Conor Ryan, Michael Tetteh, Jack McEllin, Douglas Mota Dias, Enrique Naredo, Richard Conway, Tuan-Anh Nguyen, Thi Anh-Nga Nguyen, John Charles Driscoll, Komla Agbenyo Folly, Severus Panduleni Sheetekela, Tshina Fa Mulumba, Gautam Garai

First published in London, United Kingdom, 2022 by IntechOpen
IntechOpen is the global imprint of INTECHOPEN LIMITED, registered in England and Wales, registration number: 11086078, 5 Princes Gate Court, London, SW7 2QJ, United Kingdom

British Library Cataloguing-in-Publication Data
A catalogue record for this book is available from the British Library

Additional hard and PDF copies can be obtained from orders@intechopen.com

Genetic Algorithms
Edited by Sebastián Ventura, José María Luna and José María Moyano
p. cm.
Print ISBN 978-1-80355-177-7
Online ISBN 978-1-80355-178-4
eBook (PDF) ISBN 978-1-80355-179-1

Meet the editors

Sebastian Ventura is a full professor at the University of Córdoba, Spain, where he heads the Knowledge Discovery and Intelligent Systems Research Laboratory. He received his Ph.D. in Chemistry from the same university in 1996. He has published more than 300 papers in journals and scientific conferences and edited three books and several special issues of journals. He has participated in sixteen research projects (serving as coordinator of eight of them) supported by the Spanish and Andalusian governments and the European Union. His research interests are soft computing, machine learning, and data mining.

Jose M. Luna received a Ph.D. in Computer Science from the University of Granada, Spain, in 2014. He is currently an associate professor in the Department of Computer Science and Numerical Analysis, University of Cordoba, Spain. He has published two books, two book chapters, and more than thirty articles in journals and international scientific conferences. He has also been involved in four national and regional research projects as well as three international projects. His research is focused on evolutionary computation and pattern mining.

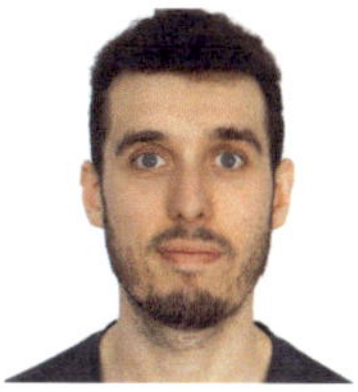

Jose M. Moyano obtained his Ph.D. in Computer Science from the University of Córdoba, Spain, and Virginia Commonwealth University, USA in 2020. He also received his BSc and MSc in Computer Science from the University of Córdoba in 2014 and 2016, respectively. Currently, he is an assistant professor at the University of Córdoba and a member of its Knowledge Discovery and Intelligent Systems Research Laboratory. To date, he has published eighteen articles in indexed journals and international scientific conferences. He has also participated in five national and regional research projects. His research interests include ensemble methods for multi-label classification.

Contents

Preface

The solution to many real-world problems lies in optimizing processes, parameters, or techniques. However, these optimizations usually mean dealing with immense search spaces and thus require exhaustive methods to evaluate all possible solutions in the search for a global optimum. In addition, many local optima may exist in the search space, so simple techniques may get stuck in them. Evolutionary algorithms and more concrete genetic algorithms are metaheuristic techniques inspired by Darwin's theory of natural selection to solve search-based optimization problems. These algorithms have been demonstrated to effectively deal with complex search spaces. Genetic algorithms employ a population of individuals, each representing a full or partial solution to the problem, bred and reproduced looking for optimal individuals. These individuals are evaluated according to a fitness function, which determines how a given individual adapts to the problem at hand.

In recent years, genetic algorithms have advanced by proposing novel algorithmic flows, representations, or specific techniques inside the main structure of the algorithm. As a result, genetic algorithms have been successfully applied to solve many real-world problems (engineering, smart cities, and energy). They have also helped to improve many machine learning (classification, regression, or hyperparameter optimization) and data mining (data preprocessing, pattern mining, or feature selection) techniques.

This book provides a comprehensive overview of the current state of the art and advances in genetic algorithms and examines the fields in which they have been applied throughout the years. It is divided into several sections, including an introduction to genetic algorithms and a summary of their applications in numerous scientific fields. An additional section includes chapters related to engineering applications, covering fields such as power systems signal stability, design of digital circuits, chemical optimization, and controller systems. This book also describes the use of genetic algorithms in architecture.

Sebastián Ventura, José María Luna and José María Moyano
University of Cordoba,
Cordoba, Spain

Section 1

Introduction

Chapter 1

Introduction to Evolutionary Algorithms

S. Tamilselvi

Abstract

Real-world has many optimization scenarios with multiple constraints and objective functions that are discontinuous, nonlinear, non-convex, and multi-modal in nature. Also, the optimization problems are multi-dimensional with mixed types of variables like integer, real, discrete, binary, and having a different range of values which demands normalization. Hence, the search space of the problem cannot be smooth. Evolutionary algorithms have started gaining attention and have been employed for computational processes to solve complex engineering problems. Because it has become an instrument for research scientists and engineers who need to apply the supremacy of the theory of evolution to shape any optimization-based research problems and articles. In this chapter, there is a comprehensive introduction to the optimization field with the state-of-the-art in evolutionary computation. Though many books have described such areas of optimization in any form as evolution strategies, genetic programming, genetic algorithms, and evolutionary programming, evolutionary algorithms, that is, evolutionary computation is remarkable for considering it to discuss in detail as a general class.

Keywords: evolutionary algorithms, genetic operators, non-convex, multi-modal, optimization process

1. Introduction

Darwin's principle of evolution says that the existence of any creature is based on the law "strongest creature survives." Before computers have entered the human world, in the 50s, knowledge to apply Darwinian principles for automated problem solving was invented. Darwin also proved that the survival of any organism can be maintained with genetic inheritance, such as reproduction, crossover, and mutation. Thus, Darwin's evolution theory was deployed by computational optimization algorithm to search for a solution to any real-world optimization problem in a natural way [1].

In the 60s, three various interpretations of this idea were introduced at different places. Evolutionary programming was developed by Lawrence J. Fogel in the USA when John Henry at Holland started his methodology as a genetic algorithm,

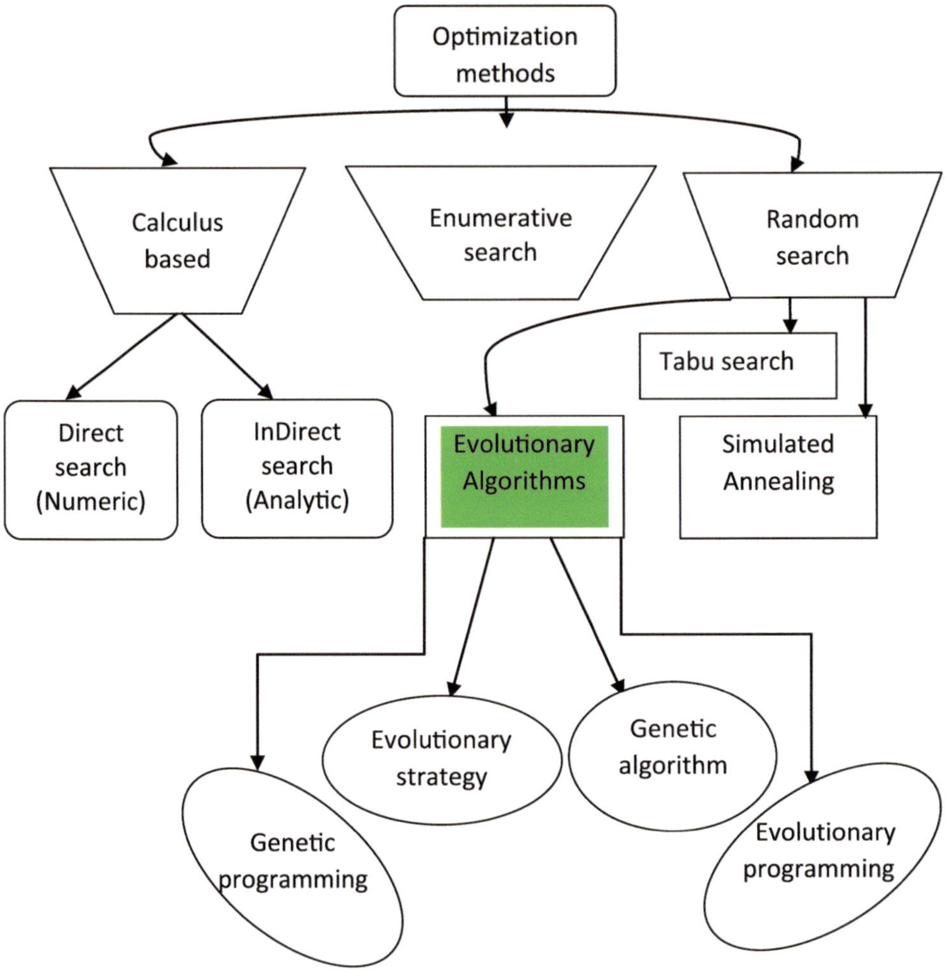

Figure 1.
Evolutionary algorithms and their subtypes.

stimulated by Darwin's evolutionary concepts. Similarly, Ingo Rechenberg and Hans-Paul Schwefel have invented evolution strategies in Germany. Following this fourth one had emerged as genetic programming, in the early 90s. These four different terminologies are seen as different representatives of one technology called evolutionary algorithms (EAs), which denote the whole field by considering evolutionary programming, evolution strategies, genetic algorithms, and genetic programming as sub-areas and is well depicted in **Figure 1** [1, 2].

2. Need for evolutionary algorithms

Real-world has many optimization scenarios. Optimization, by definition, is a methodology of making the decision as fully perfect as possible to achieve the maximum possible goal, in an engineering system. Nature is a very good optimizer. An optimization problem can be stated as follows [2, 3].

$$\text{Find } x = \{x_1, x_2, \ldots x_n\}, \text{which}$$
$$\text{Minimize/Maximize} f(x)$$
$$\text{Subject to}$$
$$g_j(x) \leq 0, j = 1, 2, \ldots, m$$
$$h_j(x) = 0, j = 1, 2, \ldots, p$$

Any engineering system can be represented with a set of quantities. Certain quantities are usually fixed called as pre-assigned constants. Remaining quantities can be treated as decision variables in the optimization process, $x_i = \{x_1, x_2, \ldots x_n\}$. '$f(x)$' is the objective function or goal to be attained, '$g_j(x)$' represent 'm' the count of inequality constraints and '$h_j(x)$' represent 'p' count of equality constraints to be satisfied for attaining feasibility [3].

In real-world engineering problems, the objective function is discontinuous, nonlinear, non-convex, and multi-modal. Also, the problems are multi-dimensional as the number of design variables are more and they are mixed in type like integer, real, discrete, binary. Hence, the search space is not smooth. It may require accessing look-up table data for objective function evaluation. The constraint functions are very complex and the amount of violation of each constraint does not cover the same range, which requires normalization [4].

In general, the optimization problems are categorized based on the existence of constraints, nature of the decision variables, permissible values of the design variables, nature of the equations involved, deterministic nature of the variables, separability of the functions, number of objective functions, etc. Some of them are static optimization problem, dynamic optimization problem, linear programming problem, convex programming problem, nonlinear programming problem, geometric programming, quadratic programming problem, separable programming problem, multi-objective optimization problem, single-variate optimization problem, multi-variate optimization problem [5].

Two different techniques to find the solution for optimization problems are mathematical programming techniques and meta-heuristic techniques. When derivative-based mathematical programming methods are applied in solving nonlinear programming problems, there are several shortcomings. Traditional optimization methods:

- Yields results that are caught at premature convergence, that is, local optima often, due to the search space with multi-modality.
- Requires mathematically well-defined objective and constraint functions.
- Requires existence of derivatives for objective function and constraint functions.
- Find difficulty to handle mixed variables.

For a real-world optimization problem, the surface plot obtained even for optimizing two design variables gives greater number of local minima/maxima. **Figure 2** depicts the surface plot obtained for the design optimization of the distribution transformer problem with two decision variables, width of the core leg, height of the core window, by minimizing the transformer lifetime cost (TLTC) of the transformer. It is

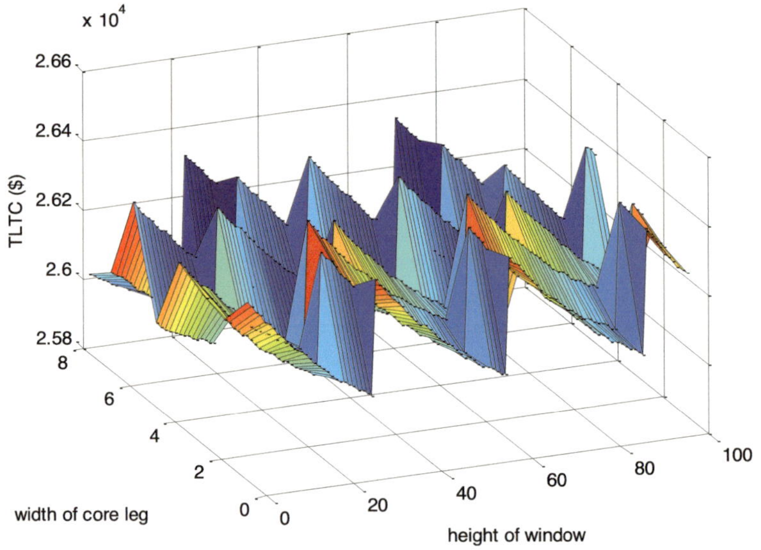

Figure 2.
Search space for minimization of TLTC objective optimizing two decision variables.

very clear from **Figure 2** that the real-world problem is very complex with its multi-modal search space [3].

Of course, when it is a multi-dimensional engineering problem, which requires optimization of more decision variables, the multi-modality cannot be imagined. So, the conventional derivative-based mathematical programming technique cannot handle such complex nature of the optimization problem, accurately. It may yield a feasible design; however, it will not be an optimal solution. If there is no knowledge or little knowledge about the behavior of the objective function related to the presence of local minima, location of feasible region, infeasible region in the multi-dimensional parameter space, it is advisable to start the meta-heuristic technique, which is a stochastic strategy [2, 3].

Of all the meta-heuristic techniques, evolutionary algorithms (EAs) are especially effective in the solution of high-complexity, non-convex, nonlinear, multi-dimensional, mixed variable, multi-objective, constrained optimization problems, for which a traditional mathematical model is difficult to build, where the nature of the input/output relationship is neither well defined nor easily computable. The stages of EAs have not yet been investigated in detail steps with illustration, despite their performances are better in terms of convergence, consistency in obtaining the solution, and computational speed in solving any multi-modal problems.

Hence, this chapter discusses in detail the step-by-step evolutionary process that happens behind the optimization algorithm. It highly helps to find solution for any multi-modal real-world engineering optimization problem, by optimizing design objective, while satisfying simultaneously various constraints imposed by international standards and user specifications.

3. Known optimization problems

Evolutionary optimization algorithms minimize or maximize an objective function and they are search algorithms. The algorithm checks all the way through a large search space of possible solution set for the optimal best solution. In day-to-day practical life as well as professional life, there are numerous activities that seek optimization. Some of the common well-known real-world optimization problems are the traveling salesman problem, classification problem, economic power dispatch, base station location problem, antenna design, scheduling problem, etc.

In traveling salesman problem, a salesman wants to visit all the towns, with information of list of towns and distances between all the towns. The constraint is that each town has to be visited only once. The optimization problem statement is searching for the optimum shortest distance/route that the salesman travel and visits each town exactly only once and returns to the place where he started [6].

Base station location problem is setting radio and optimizing maximum coverage. Given a set of spots for installing base stations, feasible configurations for every base station, antenna tilt, maximum power, antenna height, sectors orientation, etc. along with the information of traffic and strength of the signal propagation, the optimization algorithm is to choose the right location and appropriate configuration of the base station such that the installation cost is minimum while meeting the traffic demand simultaneously [7].

The job of optimal generator maintenance scheduling problem is to find out the optimum period for which, the generator units must be taken offline for maintenance over the stipulated time horizon, so that the operating costs involved are minimized, meeting the maintenance constraints during the considered time period such as load demand, maintenance window, maintenance duration and manpower availability [8].

Previous research works have applied only machine learning techniques for the prediction and classification of any disease/tumor. However, nowadays due to the capability of evolutionary algorithms, such classification problems have been stated as an optimization problem and solved using the integrated machine learning-optimization technique. Thus, optimal classification problem aims to select optimum elite features from intelligent liver and kidney cancer diagnostic systems of huge data sets, by filtering the redundant features, minimizing the error rate, in order to improve the quality of heart disease classification [9].

Economic power dispatch is a vital optimization problem in power system planning. The aim of the economic dispatch is to schedule the optimum power output for the available generator units of the power system such that the production cost is minimum and power demand is met [10].

4. Optimization process of simple evolutionary algorithm

EA handles a population of possible random solutions. Each solution is represented through a chromosome. The fitness of each chromosome is calculated to call for a competition among the chromosomes. Competition results in the selection of those better chromosomes/solutions (with high fitness value) that are suited for the environment. The process of the first level of filtration based on the fitness value is called parent selection [3]. The selected individuals, that is, parents act as seeds for creating children through genetic inheritance, that is, recombination and mutation. These genetic operators aid the necessary diversity. Few pairs of chromosomes from the parent pool are chosen based on the random probability to undergo crossover for forming offspring. The resulted offspring individuals obtained after crossover are allowed to take up mutation randomly. Different regions of the search space are explored for identifying possible optimal individuals through “recombination and mutation” operation known as “exploration.” The

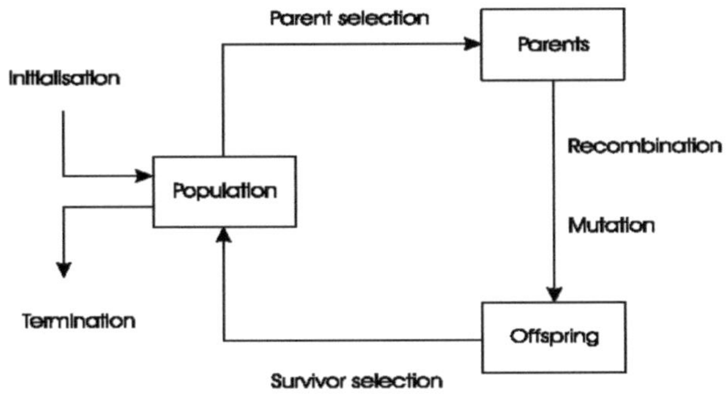

Figure 3.
Working of evolutionary algorithms.

new individuals/children thus formed have their fitness evaluated to compete for survival in the next generation. As an end to an iteration, in a replacement stage, 80% worst solutions of the initial random population are substituted by the best offspring children filtered after survival selection process based on the evaluated fitness value. Over time and several iterations, "natural selection" operation, which is called exploitation leads to the identification of an individual in the population as global optimum. The complete working of the evolutionary algorithm is pictured in **Figure 3** [11]. The major steps involved in the process of optimization in an evolutionary algorithm are as follows [1, 3].

- Solution representation
- Random population generation
- Fitness function evaluation
- Parent selection
- Reproduction—(crossover, mutation)
- Survival selection
- Replacement
- Stopping criteria

5. Iterative process behind evolutionary algorithms

To define problem statement:

Consider an equality function, $x + 2y + 3z + 4u = 30$. We shall apply the evolutionary algorithm to find the appropriate values for x, y, z, u, such that the equity equation gets satisfied [12].

5.1 Formulation of optimization problem

a. Formulate objective function: $f(k)$

The objective/aim is to **Minimize** $f(k) = [(x + 2y + 3z + 4u) - 30]$.

b. Identify decision variables/type: In this equity problem, there are four decision variables $[x, y, z, u]$. Variables that possess a larger influence on the objective function and constraint functions are appropriate ones to be chosen as decision variables.

c. Find problem dimension:

Total number of decision variables is the problem dimension = 4.

d. Representation:

A solution generated by an evolutionary algorithm is called a chromosome/individual, which is made up of genes [1]. After selecting the decision variables, and problem dimension, choice of suitable type for these variables is another important task. The nature of the decision variables is completely problem dependent and thus in this example, $[x, y, z, u]$—they are integers. However, the genes can be mixed like binary, real, integer, discrete variables, etc., depending upon the need of the problem under consideration. Chromosome/solution is thus represented as an integer variable as under:

x	y	z	U

e. Impose boundary constraint:

This range selection for setting the search space is more often done on a trial basis, in case the problem dimension is high. On contrary, if the objective function is very simple, clear, and possesses straight relationship (mathematical equation) with lesser number of decision variables, then the search space can be decided by inspection [2]. For this example, it is very clear that the integer values of decision variables $[x, y, z, u]$ can be restricted to vary between 1 and 30, in order to speed up the computational search.

5.2 Different stages in optimization process

To illustrate solving a minimization type optimization problem using EA, integer type for decision variables, six for population size, single-point method for crossover, and roulette wheel for selection are assumed. The various stages involved in the process of optimization are given for one iteration in this section [13–15].

Stage 1: *Population generation: Initial solution set*

Four genes $[x, y, z, u]$ are generated randomly satisfying the lower and upper limits of the boundary constraint. A chromosome thus generated is a vector comprising of four genes. Chromosome refers to the solution/individual of the formulated optimization problem, while the collection of such chromosomes is referred to as a population. For example, Solution [1] = $[x; y; z; u]$ = [12, 05, 23, 08]. Then, the initial population will have an array of sizes [population size, problem dimension]. That is, [(6, 4)] as shown in **Table 1**.

Stage 2: *Function evaluation: Feval*

All the chromosomes of random population will then go through a process known as fitness evaluation to measure the quality of the solution created by EA. Evaluation

	Initial random population				Feval [k]	Remarks
Solution [1]	12	05	23	08	93	
Solution [2]	02	21	18	03	80	
Solution [3]	10	04	13	14	83	
Solution [4]	20	01	10	06	46	**Best Solution**
Solution [5]	01	04	13	19	94	
Solution [6]	20	05	17	01	55	

Table 1.
Functional evaluation of initial population.

of fitness value of chromosome/solution is carried out by calculating the objective function value as Feval = Modulus[$f(k)$].

$$f(k) = [(x + 2y + 3z + 4u) - 30].$$

$$\text{For Solution [1], Feval [1]} = \text{Modulus [f(Solution [1])]}$$
$$= \text{mod } [(12 + 2 \text{ x } 5 + 3 \text{ x } 23 + 4 \text{ x } 8) - 30]$$

Feval [1] = mod [(12 + 10 + 69 + 32) − 30] = 93. Similarly, the solutions of entire population can be calculated and tabulated as under in **Table 1**. It is found that Chromosome 4 has the least objective function value 46.

Stage 3: *Parent selection*

The chromosomes are selected from the initial population to act as parent for reproduction, based on the fitness of the solution/individual. The selection procedure tells how to choose individuals in the population that will create offspring for the next generation. The fittest solution will have a higher probability to be selected as a parent. Two-step selection process is discussed as follows [13, 15].

A. To compute the probability of selection: Prob[i]

$$\text{Prob}[i] = \frac{\text{Fit}[i]}{\sum_{i=1}^{6} \text{Fit}[i]}$$

where,

$$\text{Fit}[i] = \frac{1}{(1 + \text{Feval}[i])}$$

(to avoid undefined divide by zero error, which may encounter for the optimal solution, it is advisable to add 1 with Feval).

Fit [1] = 1/(1 + Feval [1]) = 1/94 = 0.0106 and so on, till i = 6.

Total fitness = 0.0845 (refer **Table 2**).

Prob [1] = 0.0106/0.0845 = 0.1254

B. To select the parent pool: Roulette-wheel (RW) selection process:

Parent selection is vital for the convergence of optimization algorithm as efficient parents force solutions/individuals to optimality. There are different methods in the

Initial population				Feval [*i*]	Fit [*i*]	Prob [*i*]	Cum [*i*]
12	05	23	08	93	0.0106	0.1254	0.1254
02	21	18	03	80	0.0123	0.1456	0.2710
10	04	13	14	83	0.0119	0.1408	0.4118
20	01	10	06	46	**0.0213**	**0.2521**	0.6639
01	04	13	19	94	0.0105	0.1243	0.7882
20	05	17	01	55	0.0179	0.2118	1
		Total fitness			0.0845		

Table 2.
Selection probability computation.

process of selecting parents such as stochastic universal sampling, fitness proportionate selection, tournament selection, rank selection, and random selection. In this chapter, roulette wheel selection has been implemented for identifying the right parent pool.

Consider a wheel that is split into '6' pies. Pie refers to the individual in the population. Each solution occupies a portion of the wheel, proportional to its fitness value. It is clear that a fitter solution takes a larger pie on the wheel and has larger probability chance of being selected as a parent when the wheel is made to spin '6' times. Hence, the probability of choosing a chromosome depends on its fitness only. The steps involved in the roulette wheel selection process are:

- Compute cumulative probability values for all the solutions—Cum[i].

- Allot pie in sequence for all the '6' individuals, based on the cumulative probability. That is, Chromosome 1 has occupied light blue pie with cumulative probability ranging between [0–0.1254]. Chromosome 4 which has the highest fitness value '0.0213' has the highest probability '0.2521' among all the solutions of the population. Naturally, it will take larger sized pie, which is yellow in color on the wheel. It is clearly explained in **Figure 4**, and **Tables 3** and **4**.

- To arrange the order of chromosomes (6) in the parent pool, equivalent to spinning the wheel '6' times, generate random number '6' times, 'rand[i]' < 1, six.

  rand [place 1] = 0.2; rand [place 2] = 0.285; rand [place 3] = 0.098;

  rand [place 4] = 0.812; rand [place 5] = 0.397; rand [place 6] = 0.50.

- Fit the random number of each place in the respective range of cumulative probabilities and fetch the color of the pie. For example, rand [place 2] = 0.285, which is between Cum [2] and Cum [3]. So, the gray pie which is individual/chromosome [3] will occur in place2 of the parent/mating pool.

Stage 4: *Crossover*

The crossover operation involves three steps: (A) selecting mating chromosomes, (B) determining cut point for crossover, and (C) updating the population.

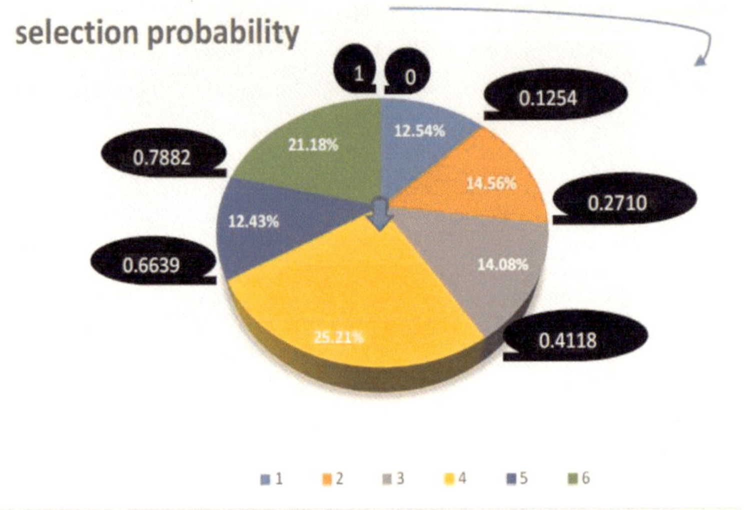

Figure 4.
Roulette wheel—Parent selection process.

Position	Chromosome		Initial population			
Place 1	I	Solution [1]	12	05	23	08
Place 2	II	Solution [2]	02	21	18	03
Place 3	III	Solution [3]	10	04	13	14
Place 4	IV	Solution [4]	20	01	10	06
Place 5	V	Solution [5]	01	04	13	19
Place 6	VI	Solution [6]	20	05	17	01

Table 3.
Place and position of solution—Before RW selection process.

In this operation pairs of parents are chosen and many children/off-springs are generated using the information available in the gene of the parents. Usually, crossover operation is deployed in EA with high probability(p_c). Some of the commonly used crossover operators are whole arithmetic recombination, one point crossover, uniform crossover, multi-point crossover, Davis' order crossover, etc. In the equity problem chosen for illustration, one-point crossover has been used for offspring creation. In one-point crossover, a randomly point of crossover has been chosen and the tail ends of the parent pairs are swapped to produce new children. The process is evident in **Table 5**.

A. To select chromosome:

Parent chromosome from parent pool that undergoes the mating process is randomly selected and the number of mate solutions is decided using crossover rate, p_c. Solution 'i' will become a parent, if random number, rand[i]falls below the crossover rate. Let us assume the $p_c = 25\%$ for solving the problem. Generate number randomly '6' times (population size) below 1.

Position	Chromosome	Population after selection			
Place 1	II	02	21	18	03
Place 2	III	10	04	13	14
Place 3	I	12	05	23	08
Place 4	VI	20	05	17	01
Place 5	III	10	04	13	14
Place 6	IV	20	01	10	06

Table 4.
Chromosomes in the MATING POOL after RW selection process.

Population after selection				Population after crossover			
02	21	18	03	02	05	17	01
10	04	13	14	10	04	13	14
12	05	23	08	12	05	23	08
20	05	17	01	20	04	13	14
10	04	13	14	10	04	18	03
20	01	10	06	20	01	10	06

Table 5.
Population after and before crossover operation.

rand [1] = 0.19; rand [2] = 0.249; rand [3] = 0.750; rand [4] = 0.005
rand [5] = 0.149; rand [6] = 0.320

Thus, for the generated random numbers, three chromosomes/solutions [1, 4, 5] are selected for crossover operation. Hence, the number of crossovers becomes 3, that is, 3 pairs.

Solution [1] >< Solution [4] — First Crossover
Solution [4] >< Solution [5] — Second Crossover
Solution [5] >< Solution [1] — Third Crossover

B. To determine cut point:

Followed by mating chromosome selection, the next phase is to determine the position of the crossover point. The steps involved are:

- Generate random numbers between 1 to (Problem dimension—1) in order to get the crossover point. That is, between 1 and 3. Assume, Cut [1] = 1; Cut [2] = 1; Cut [3] = 2
- Parent individuals get cut at the crossover point and their genes are interchanged. For first, second, and third crossovers, parents' genes are cut at positions 1, 1, and 2, respectively.

First Crossover: New Chromosome [1] = Solution [1] >< Solution [4]

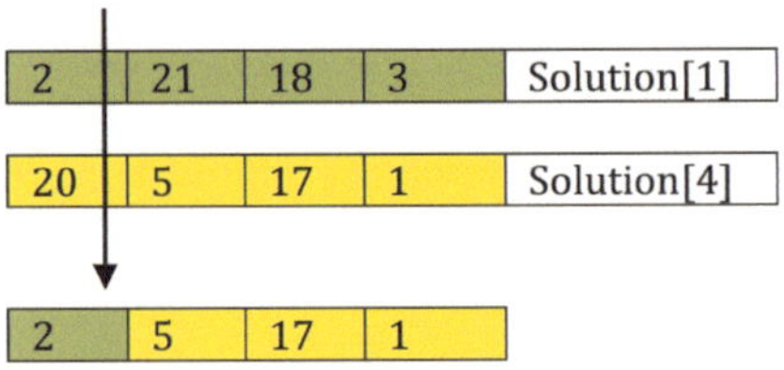

Second Crossover: New Chromosome [4] = Solution [4] >< Solution [5]

Third Crossover: New Chromosome [5] = Solution [5] >< Solution [1]

C. To update the population after crossover:

Stage 5: *Mutation*

Mutation operation is a small random sharp change in the chromosome necessary to obtain a new solution. Mutation is used to sustain population diversity and it generally has a low probability, p_m. Mutation in EA refers "exploration" of search space. It has been proven that mutation operation is crucial for the convergence of the algorithm, however, crossover operation is not so. Some of the commonly used mutation operators are bit flip mutation, swap mutation, scramble mutation, random resetting, inversion mutation, etc. Like the crossover operators, this is not an extensive list since EA designer may deploy a hybrid approach as a combination of these operators or prefer problem-specific mutation operators as more practical.

- Calculate the total number of genes in the population = 24 genes.

- Calculate number of mutable genes.

Number of solutions that undergo mutations in a population is decided by mutation rate p_m. Let, ρ_m = 10%; Number of mutations = $0.1 \times 24 = 2.4 = 2$.

- Calculate gene positions.

Generate two random numbers below 24, say 12 and 18. Mutable genes and chromosomes are Chromosome [3]-gene 4 and Chromosome [5]-gene 2. This process is seen in **Table 6**.

- The value of mutable genes at the mutation point is substituted with random number, satisfying the boundary constraint of decision variables/genes. That is, between 0 and 30; Say 02, 05.

Stage 6: *Survival selection and replacement mechanism*

After mutation operation one iteration/generation of EA is over. Functional evaluation is again performed on the offspring for survival selection. From the functional evaluation of population after mutation, it is evident that the objective function value

Population after selection			Population after crossover					Population after mutation			
02	21	18	03	02	05	17	01	02	05	17	01
10	04	13	14	10	04	13	14	10	04	13	14
12	05	23	08	12	05	23	08	12	05	23	02
20	05	17	01	20	04	13	14	20	04	13	14
10	04	13	14	10	04	18	03	10	05	18	03
20	01	10	06	20	01	10	06	20	01	10	06

Table 6.
Population after and before mutation operation.

of the best solution is reducing—37 in comparison with the minimum objective value —47 of initial random population, as shown in the table. Hence the minimization objective, $f(k) = [(x + 2y + 3z + 4u) - 30]$ is met. This means that the solutions obtained by EA at the end of the first iteration is better than the solutions of random population.

To execute the iteration process continuously, population is to be revised at the end of each iteration, as a final process, which is referred to as replacement mechanism. In each iteration end, 80–90% of best solutions from offspring population (4–5 best children) and 20–10% best solutions from the initial population (2–1 random solution) are selected to form new population for next generation [15]. Chromosomes of the next generation will then become as shown in **Tables 7** and **8**.

Population after mutation				Feval	Remarks
02	05	17	01	**37**	**Best Solution** and survive in the next generation
10	04	13	14	77	Survive in next generation
12	05	23	02	47	Survive in next generation
20	04	13	14	93	Rejected solution
10	05	18	03	56	Survive in next generation
20	01	10	06	46	Survive in next generation

Table 7.
Survival selection.

Population after mutation				Next generation initial population				Feval	Remarks
02	05	17	01	02	05	17	01	**37**	
10	04	13	14	10	04	13	14	77	
12	05	23	02	12	05	23	02	47	
20	04	13	14	20	01	10	06	47	Replaced solution
10	05	18	03	10	05	18	03	56	
20	01	10	06	20	01	10	06	46	

Table 8.
Replacement-population for the next iteration.

Stage 7: *Stopping the iteration*

The optimization process is repeated until when objective function value or decision variables values become stagnant, that is, have no/very little change for a greater number of iterations. Thus, over period, the solution will get converge to the final best minimum optimal solution and the optimization process will be stopped, based on any stopping criteria such as the maximum number of iterations, or maximum number of functional evaluations, etc.

6. Conclusion

The basic processes that occur behind an evolutionary algorithm have been explained and illustrated in this chapter with steps covering solution representation, population generation, functional evaluation, parent selection, genetic operations, offspring evaluations, survival selection, and stopping criteria for a simple optimization problem. This knowledge can be extended very well by researchers across any discipline, working in the field of optimization and for applying evolutionary algorithms to solve any complex engineering problem using computers. Although the process behind EA may appear to be simple, the details of the optimization process form the base and are very much necessary in applying the learned concepts for modifying the existing evolutionary concepts and evolving into better optimization methods in the research level.

Author details

S. Tamilselvi
Sri Sivasubramaniya Nadar College of Engineering, Chennai, India

*Address all correspondence to: tamilselvis@ssn.edu.in

References

[1] Michalewicz Z. Genetic Algorithms + Data Structures = Evolution Programs. 3rd ed. Berlin, Heidelberg: Springer; 1996. p. 387. DOI: 10.1007/978-3-662-03315-9

[2] Gen M, Cheng R. GA & Engineering Design. Hoboken, New Jersey, United States: John Willey & Sons, Inc.; 1997

[3] Tamilselvi S, Baskar S, Anandapadmanaban L, Kadhar K, Varshini PR. Chaos-assisted multiobjective evolutionary algorithm to the design of transformer. Soft Computing. 2017;**21**(19):5675-5692. DOI: 10.1007/s00500-016-2145-7

[4] Tamilselvi S, Baskar S, Anandapadmanaban L, Karthikeyan V, Rajasekar S. Multi objective evolutionary algorithm for designing energy efficient distribution transformers. Swarm and Evolutionary Computation. 2018;**1**(42): 109-124

[5] Tamilselvi S, Baskar S, Sivakumar T, Anandapadmanaban L. Evolutionary algorithm-based design optimization for right choice of transformer conductor material and stepped core. Electrical Engineering. 2019;**101**(1):259-277

[6] Agatz N, Bouman P, Schmidt M. Optimization approaches for the traveling salesman problem with drone. Transportation Science. 2018;**4**(52): 965-981

[7] Lakshminarasimman N, Baskar S, Alphones A, Willjuice Iruthayarajan M. Evolutionary multiobjective optimization of cellular base station locations using modified NSGA-II. Wireless Networks. 2011;**17**(3):597-609

[8] Tamil Selvi S, Baskar S, Rajasekar S. An Intelligent Approach Based on Metaheuristic for Generator Maintenance Scheduling—Classical and Recent Aspects of Power System Optimization. 1st ed. Cambridge, Massachusetts, United States: Academic Press; 2018. pp. 99-136

[9] Gunasundari S, Janakiraman S, Meenambal S. Multiswarm heterogeneous binary PSO using Win-Win approach for improved feature selection in liver and kidney disease diagnosis. Computerized Medical Imaging and Graphics. 2018;**70**:135-154

[10] Bhattacharya A, Chattopadhyay PK. Solving complex economic load dispatch problems using biogeography-based optimization. Expert Systems with Applications. 2010;**37**(5):3605-3615

[11] De Jong KA. Evolutionary Computation—A Unified Approach. 1st ed. Berlin/Heidelberg, Germany: Springer; 2017. p. 268

[12] Hermawanto D. Genetic Algorithm for Solving Simple Mathematical Equality Problem. 2013 [arXiv preprint arXiv:1308.4675]

[13] Available from: https://www.tutorialspoint.com/genetic_algorithms/genetic_algorithms_parent_selection.htm

[14] Goldberg DE. Genetic Algorithms in Search, Optimization, and Machine Learning. 13th ed. Boston, Massachusetts, United States: Addison Wesley; 1989. p. 432

[15] Hancock PJ. An empirical comparison of selection methods in evolutionary algorithms. In: AISB Workshop on Evolutionary Computing 1994 Apr 11. Berlin, Heidelberg: Springer; 1994. pp. 80-94

Chapter 2

Application of Genetic Algorithm in Numerous Scientific Fields

Gautam Garai

Abstract

The genetic algorithm (GA) and its variants have been used in a wide variety of fields by the scientists efficiently for solving problems. From the pool of evolutionary algorithms, the GA is chosen by the researchers and has been popular as a useful and effective optimizer. It has several advantages and disadvantages. However, it provides solutions for various kinds of problems such as space research, economics, market study, geography, remote sensing, agriculture, data mining, cancer detection, and many more. This chapter discusses the utilization of the GA in some of these fields with a few experimental results such as data clustering, pattern identification and matching, and shape detection. The results are illustrated and explained with reasons for better understanding of the GA application in the scientific fields. Other than these, the GA in bioinformatics for biological sequence alignment is discussed with examples.

Keywords: genetic algorithm, optimization, application, scientific field, stochastic search

1. Introduction

In the designing process of scientific problems, the primary goal of the researchers is to develop a system to provide an efficient low-cost solution. In this endeavor, they need a good optimizer to reach their target. Several optimizers are available, which work efficiently for various kinds of problems. Among these, genetic algorithm (GA) is chosen as a functional tool for good and useful optimizer. The algorithm has proven to be a class of effective optimization techniques for many applications in engineering, economics, manufacturing, artificial intelligence, operations research, space science, agriculture, physics, chemistry, bioinformatics, medical science, and many more. However, there are advantages and disadvantages of using genetic algorithms as an optimizer such as other optimization tools. The amount of *a priori* knowledge required to use genetic algorithms is minimal. One can apply an "off the shelf" genetic algorithm to an optimization problem by mapping all of the function inputs to a pre-specified representation, such as binary strings. However, one often specializes a genetic algorithm to a specific problem domain by using additional heuristics, specialized representation, and/or operators, which constitute the *a priori* information for obtaining superior performance. If one opts to use a genetic algorithm without expending effort to customize it to the problem at hand, the trade-off is typically that the non-specialized genetic algorithm may produce poor results, which may include computationally inefficient optimization resulting to converge to a suboptimal solution. The scientists have incorporated customization in the conventional GA according to the need and specification of their research problems.

This chapter discusses the utilization of GA in various fields with examples and illustrations. It is shown how the GA recognizes a specific pattern from a group of patterns. Genetic-algorithm-based clustering technique is then described for identifying similar group of items among several homogeneous or heterogeneous groups. The discussion is also done on the biological sequence alignment with the GA in achieving accurate alignment.

The rest of the chapter is organized as follows. Section 2 describes briefly the history of evolutionary algorithm along with the GA. A survey on genetic algorithm is narrated in Section 3. The next three sections elaborate several utilizations of the GA in various scientific fields. Section 4 identifies a pattern among a group of patterns. Genetically guided clustering technique is elaborated in Section 5. Section 6 describes the application of genetic algorithm for detection of polygonal shape of a dot pattern. The biological sequence alignment is discussed in Section 7. Section 8 concludes the chapter.

2. A brief history of evolutionary computation

In the literatures, three primary search methods are identified, namely *calculus-based*, *enumerative,* and *random* [1] search. *Calculus-based* methods are further subdivided into two classes—*indirect* and *direct*. Indirect search method seeks the local extrema by solving a set of equations resulting from the derivative of objective function. On the other hand, direct search method seeks the local optima by hopping on the function and moving in a direction related to the local gradient. This is basically the notion of *hill-climbing*. *Enumerative* scheme has been considered in many shapes and sizes. Here, the search algorithm finds the objective function value at every point within a finite or a discrete infinite space. *Random* search is the most popular among the researchers since the shortcomings of other two techniques are rectified here. The genetic algorithm is an example of a search procedure of this category [1].

For several years since 1950, many computer researchers studied evolutionary systems independently with an idea that in future evolution could be used as an important and essential optimization tool for solving various problems in engineering. The idea was to develop a system with a population of candidate solutions to a given problem with operators influenced by both natural selection and natural genetic variation. Rechenberg [2] introduced "evolution strategies" (*Evolutionsstrategie* in the original German), a method used to optimize real-valued parameters for devices such as air foils. This idea was further developed by Schwefel [3]. Fogel et al. [4] developed a technique called "evolutionary programming" where the candidate solutions to the given tasks were represented as finite-state machines, which were evolved by randomly mutating their state-transition diagrams and selecting the fittest among them. The evolution strategies, evolutionary programming and genetic algorithms together form the backbone of evolutionary computation. Several other researchers developed evolution-inspired algorithms for optimization and machine learning. Box [5], Friedman [6], Bremermann [7], and Reed et al. [8] worked in this area though their studies had little impact on the field. Evolutionary biologists also used computers to simulate evolution for the controlled experiments [9–11].

Finally, in 1960s and 1970s, at the University of Michigan, Holland first invented the genetic algorithms and later on improved by himself along with his colleagues [1]. Unlike the previous evolution strategies, Holland's intention was to formally study the phenomenon of adaptation, which occurs in nature instead of designing algorithms for solving specific problems. He also developed ways by importing the natural adaptation mechanisms in computer systems. The genetic algorithm was presented by Holland as an abstraction of the biological evolution and also gave a theoretical framework for adaptation. His GA was a method for moving from one population of "chromosome" (string of 1's and 0's) to a new population with a kind of "natural selection" along with the genetic operators, namely crossover, mutation, and inversion. Each chromosome was a combination of "genes" (bits), and each gene was a particular "allele" (either 0 or 1). In the population the useful chromosomes for reproduction were chosen by the selection operator. The fitter chromosomes thus produced more off-springs compared with less fit ones. In the process of crossover, the subparts of two chromosomes were exchanged. This basically mimicked the biological recombination between two single-chromosome (haploid) organisms. The allele values of some locations of the chromosome were randomly changed and reversed by mutation and inversion processes, respectively.

For last several years, the researchers had been studying and interacting widely on different evolutionary computation methods and ultimately it had broken down to some extent the boundaries between GAs, evolutionary programming, evolution strategies, and other evolutionary approaches. Today the term "genetic algorithm" is used by the researchers to describe something very far from the original conception of Holland. There are at least three overlapping meanings of *search* in genetic algorithms and other search procedures, as discussed below. *Search for stored data* – Here the problem is to efficiently retrieve information stored in the computer memory. Suppose in a large database of names and addresses stored in some ordered way we want to search for a record corresponding to a given last name using *Binary search* procedure. *Search for paths to goal* – Here the problem efficiently searches to find a set of actions that will reach a given goal starting from a given initial state. This form of search is central to many techniques in artificial intelligence. **Figure 1** illustrates a simple example of *8-puzzle*. A set of tiles numbered 1–8 are placed in a square, leaving one space empty. Sliding one of the adjacent tiles into the blank space is termed a *move*. Typical algorithms are *branch and bound search*, *depth-first search,* etc. *Search for solutions* – This is rather a general class of search compared with *search for paths to goal*. Here, the concept is to efficiently search a final solution to a problem in a huge space of candidate solutions. Among these solutions, one may or may not be the goal and the solution may or may not be improved with the progress of the search. These are the kinds of search problems for where the genetic algorithms are used.

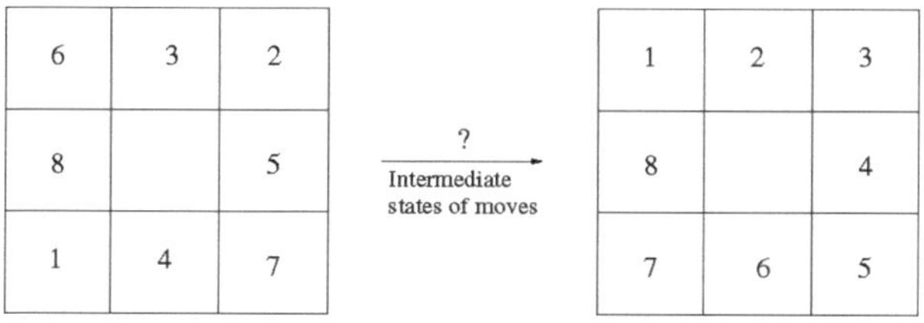

Figure 1.
Eight-puzzle problem with the initial state and the final state.

Although the genetic algorithm is directed to *search for paths to goal* or *search for solutions*, it cannot be guaranteed that it will always reach the goal. This is due to its stochastic or random search nature. On the other hand, this characteristic enables the rapid determination of a solution. With the advancement of the search process, GAs follow certain specific steps, which are not similar to the traditional search procedures.

3. A survey on genetic algorithms

The GAs have been employed in a wide variety of problems. Some of the studies involving medical image registration, image segmentation, and contour recognition are available in [12–14]. In addition, the classification of endothelial cells in tissue section, normalization of Chinese handwriting, and evaluation of earthquake risk for geological structures are examples of some practical applications [15–17]. GAs have also been used in optimization of feature extraction chain, error-correcting graph isomorphism, and dot pattern matching [18–20]. Moreover, the techniques to generate fuzzy rules for target tracking, constrained placement problems, process planning for job shop matching and blind channel identification with higher-order cumulation fitting have employed GA [21–24]. In the field of information retrieval, GA is used to retrieve the dynamic web-based contents [25]. Raidl et al. worked with biased mutation in evolutionary algorithm for solving subset-selection problems on complete graphs [26].

Although the simple genetic algorithm has been used for solving various problems, researchers have always tried to enhance the performance of GA by modifying the algorithm. Among several approaches for the improvement of performance, in 1970, Cavicchio developed a technique to conserve the best individuals by substituting the inferior parents if the offspring's fitness exceeded that of the inferior parent [27]. The *crowding* scheme of De Jong [28] intended to retain the diversity and the best individuals in the population by exchanging the maximally similar strings. Fogel [29] as well as Back and Schwefel [30] tested some optimizing functions for their algorithms and showed how the evolutionary method with self-adaptive mutation performs better than the method without self-adaptive mutation. Other than simple genetic approaches, Davis suggested that hybridizing genetic algorithms with the most successful optimization methods for particular problems provided the best performance [31]. The hybrid genetic scheme is aimed to hill-climbing from several points scattered in the search space. In case of multimodal objective function, it is obvious that some chromosomes/strings (offspring) will be in the basin of attraction of the global solution, where hill-climbing is an effective and fast form of search. However, a hybrid approach spoils hyperplane sampling, but does not destroy it entirely.

Some more variants of GA are also found in the literature [1, 32–35]. However, the implementation of modified genetic algorithm often increases the computation time for solving various complex problems, and the researchers have tried to increase the speed of the algorithm using parallel/distributed GA when the computation time is large for a particular problem. Various parallel implementations of GAs are discussed in [36, 37]. Multiobjective optimization problem is also solved by parallel genetic algorithm. The major parallel multiobjective genetic algorithms are discussed and some observations are included in [38]. Furthermore, two approaches of parallel GAs, namely the *inland* model and the *diffusion* model, are discussed in [39, 40], respectively. In the inland model, the population is divided into a number of smaller populations. Among the subpopulations, the migration of individuals happens occasionally during the progress of search process. However, the selection of individuals for migration and the frequency of migration are significant debatable problems [36]. On the contrary, each individual in the diffusion model is related to a spatial location on a low-dimensional grid. The entire population is regarded as a group of active individuals, which interact only with the neighbors. Another important utilization of distributed GA is noticed in the performance-driven VLSI routing problem, which can handle both electrical and topological constraints of the problem [41].

Other than these, the well-known NP-hard Traveling Salesman problem on a cluster of nodes is also solved by the application of parallel genetic algorithm [42]. A *master-slave* technique is implemented in the parallel/distributed genetic approach for obtaining the optimal and/or suboptimal traveling path(s). Moreover, the distributed genetic algorithm is employed in the labor scheduling problems to ascertain the number of employees and their work schedule. The objective is to minimize the labor expenses and expected opportunity costs [43].

The researchers have also successfully employed GAs in solving various kinds of problems in other scientific fields. Notredome and Higgins [44] developed a popular software for aligning sequences with two objective functions. Several GA-based approaches were developed for solving multisequence alignment [45–49]. In chemistry, GAs were incorporated in studying water oxidation [50], magnetic storage [51], catalysis [52], stability of boron wheels [53]. Sathya et al. developed a genetic-based algorithm to classify genes for identification of biomaker genes to suggest an individualized treatment [54]. As a robust heuristic search method, the GA helped for mining information from the large datasets. It was applied to discover interesting and useful relations between data elements with abstraction in the domain of association rules [55]. Another GA-based classification technique was used to improve the feature selection with support vector machine (SVM) classifier. Feature selection was used to classify a breast cancer dataset with 699 instances into two classes, namely benign and malignant, each with 11 attributes [56]. A genetic approach was also applied in machine learning for selecting the proper and the best move in playing of chess. The technique helped in deciding the effective move by classifying the set of rules for a particular solution [57]. In addition, several GAs and hybrid GAs were proposed by the medical practitioners as well as the scientists to detect cancer by different approaches such as feature selection, categorization, and classification, registering temperature difference, and many more [58]. A hybrid genetic approach was developed for early diagnosis of breast cancer using thermography. This technique measured the temperature difference between cancerous and healthy tissues with a high success rate [59].

In the following sections, some applications of genetic algorithm with several illustrations are discussed. The parameters of experimental setup are provided to help implementation of the GAs in solving similar or different kinds of problems. Some exceptional results are also elaborated with illustrations to understand the strength of GA for finding solution in numerous scientific fields.

4. Dot pattern matching identification

In a 2-D or 3-D feature space, a dot pattern represents some physical objects or class of objects with a set of dots or points. Such dot patterns are used in geographic and cartographic data, spatial information system, astronomy and astrophysics, remote sensing, image texture analysis, biomedical imaging, and some other areas [60–63]. The involvement of such studies with dot patterns includes the set estimation and shape identification, identification of point process parameter, and clustering and classification.

The dot pattern matching problem is defined as follows. T is a given test dot pattern, which is to be identified in an unknown scene of dot patterns S (a set of dot patterns or objects). Now, to find the position of best match in S, T should be translated and rotated. So a reference point (the centroid of the coordinates of the dots of T) is required to translate T. Let this centroid be O. In the space S, the translation of dot pattern to a point P indicates that their centroid is translated to P. Similarly, its rotation by θ is the rotation of pattern(s) with respect to O as origin.

The matching score is now evaluated as follows for two dot patterns S and T. Once T is transformed and rotated with respect to the origin O, the distance $d(t_i, s_j)$ between a point t_i of T and each of the points s_j of S is calculated and the minimum distance is considered. If the number of points of T is α, then the sum of the minimum distances D_{min} for all α points is as follows.

$$D_{min} = \sum_{i=1}^{\alpha} Min\left[d(t_i, s_j)\right]_{j=1,2,\dots,\beta} \tag{1}$$

where $s_1, s_2, \dots s_\beta$ and $t_1, t_2, \dots t_\alpha$ are the points of S and T, respectively, and $\alpha \le \beta$. D_{min} indicates the best matching score of two dot patterns or objects.

In pattern recognition problem, the performance of CGA (conventional genetic algorithm) [1] has been tested with the sequential and distribution of two GA-based modified methods, namely, cascaded genetic algorithm (CAGA) [64] and distributed CAGA (DCAGA) [65].

In dot pattern or object matching problem, S is a set of dot patterns of different shapes as depicted in **Figures 2** and **3**, and we have taken one of them as a test pattern T and matched it with S. The scores for matching between T and S for the genetic methods are shown in **Tables 1–3**. Here the score for successful matching is denoted by a ratio of two numbers. One of them is the number of times the solution has been reached and the total number of trials in percentage. Two types of matching, namely

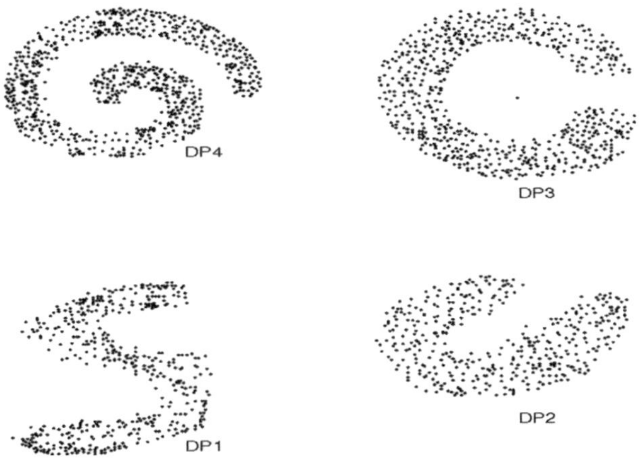

Figure 2.
A scene of multiple dot patterns in 2-D space [64].

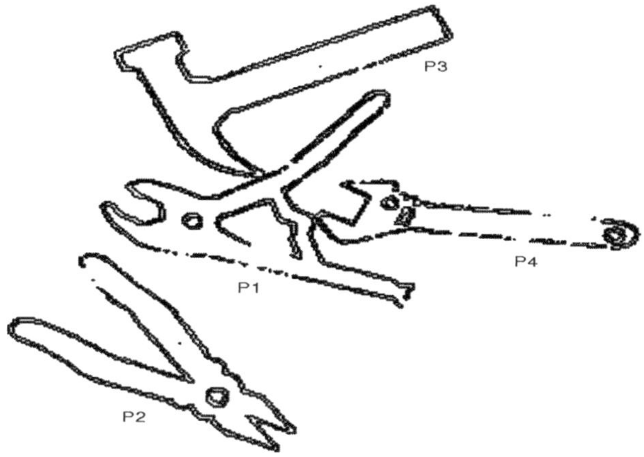

Figure 3.
A scene of multiple objects with edge map in 2-D space [64].

Dot pattern	CAGA			CGA		
	Successful matching		Unsuccessful matching	Successful matching		Unsuccessful matching
	Perfectly matched	Visually matched	Not matched	Perfectly matched	Visually matched	Not matched
DP1	22%	53%	25%	0%	10%	90%
DP2	30%	40%	30%	0%	7%	93%
DP3	73%	23%	4%	3%	53%	44%
DP4	17%	80%	3%	0%	50%	50%

Table 1.
*Matching results of CAGA and CGA on the dot patterns in **Figure** 2. The following data have been summarized over 50 runs for each DP.*

Dot pattern	DCAGA			DCGA		
	Successful matching		Unsuccessful matching	Successful matching		Unsuccessful matching
	Perfectly matched	Visually matched	Not matched	Perfectly matched	Visually matched	Not matched
DP1	57%	43%	0%	14%	73%	13%
DP2	40%	60%	0%	10%	70%	20%
DP3	83%	17%	0%	10%	67%	23%
DP4	40%	60%	0%	10%	90%	0%

Table 2.
*Matching results of DCAGA and DCGA on the dot patterns in **Figure** 2. The following data have been summarized over 50 runs for each DP.*

Dot pattern	DCAGA			DCGA		
	Successful matching		Unsuccessful matching	Successful matching		Unsuccessful matching
	Perfectly matched	Visually matched	Not matched	Perfectly matched	Visually matched	Not matched
P1	33%	67%	0%	0%	30%	70%
P2	76%	24%	0%	10%	63%	27%
P3	27%	67%	6%	3%	10%	87%
P4	17%	80%	3%	0%	27%	73%

Table 3.
Matching results of DCAGA and DCGA on the objects with edge map in **Figure 3**. *The following data have been summarized over 50 runs for each DP.*

perfect matching and *imperfect* but *visual matching*, are considered. In case of *visually matched patterns*, it is observed that *T* is superimposed over *S* without any visible error, but the matching error is not negligible. On the contrary, the calculated error of matching between *T* and *S* is close to zero in *perfectly matched situation*. Naturally, the pattern or object is termed as visually matched.

The dot/point patterns depicted in **Figure 3** are basically the *gray-tone* digital images. The pattern recognition experiment performed over these patterns is matching of edge maps. In **Figure 3,** the *gray-tone* digital image is a combination of four different objects of *512X512* pixels with *256* possible gray levels. The *Sobel Operator* is used to convert the digital image (**Figure 3**) into a two-tone edge map [66]. The edge map is usually considered as a discrete dot pattern with a dot represented digitally by 1 and a blank (white space) digitally by 0. In **Figure 3**, the object (P2) is separated from other three objects (P1, P3, and P4), which are very close and touched each other.

We consider the results of GAs for **Figure 2** in the experiment. From **Table 1**, the success rate of CAGA for the best case is 96% for DP3 where 73% of times the pattern is perfectly matched and 23% of times it is visually matched. The success rate of CAGA for DP4 is 97%, but in this case the pattern is perfectly matched for only 17% of times. The performance of CAGA is worst for DP2 with failure rate of 30%. On the other hand, the best performance of CGA is achieved for DP3 where the pattern is perfectly matched for 3% of times and visually matched for 53% of times. In CGA, the successful matching score of DP2 is only 7% (visually matched), which is worst among all four patterns.

From **Table 2**, it is noted that for the DCAGA, the success rate is 100% for all four patterns considering perfect and visual matching. The DCAGA achieves the best performance for DP3. On the contrary, in the worst case, the failure rate of the DCGA is at most 23%. However, for matching of the pattern DP4, a success rate of 100% is reached by the DCGA.

Table 3 tabulates the experimental results of object matching for the GA-based distributed approaches. It is observed that the DCAGA cannot always achieve the success rate of 100% as noticed for the earlier experiment of dot pattern matching in **Table 2**. The success rate of the DCAGA is 100% for P1 and P2, and the failure rate is a maximum 6% for P3 and P4. It is noted that both techniques perform best for the isolated object P2. The DCGA does not perform well for the objects, which are close to each other.

5. Genetically guided clustering algorithm

In general, the clustering problem can be presented as follows:

For the set of n data $\mathcal{X} = \{\mathbf{x}_1, \mathbf{x}_2, \dots, \mathbf{x}_n\}$ to be clustered, each $\mathbf{x}_i \in \mathcal{R}^p$ is an attribute vector consisting of p real measurement describing the i-th object. The objects corresponding to the data are to be clustered into non-overlapping groups $C = \{C_1, C_2, \dots, C_k\}$ where k is number of clusters, $C_1 \cup C_2 \cup \dots \cup C_k = \mathcal{X}$, $C_i \neq \phi$, and $C_i \cap C_j = \phi$ for $i \neq j$. The objects within each group should be more similar to each other than the objects in any other group, and the value of k may or may not be known.

Clustering approaches are semi-optimal ways of arriving at the grouping problem. The number of ways of sorting n objects into k groups is enormous, which is given as

$$N(n, k) = \frac{1}{k!} \sum_{i=0}^{k} (-1)^i \binom{k}{i} (k - i)^n \tag{2}$$

For 25 objects with five clusters, this is in the order of 10^{16}. Clearly, it is impractical to exhaustively search the solution space and obtain the optimal solution.

5.1 Measures of similarity

Since similarity is fundamental to the definition of a cluster, a measure of similarity between two objects/patterns/entities drawn from the same feature space is essential to most clustering procedures. The proximity of individuals (i.e., objects) is usually expressed as a distance during the clustering of data units. The clustering of variables generally involves a correlation or other such measure of association. The smaller the distance between the objects, the higher is the similarity. Because of the variety of feature types and scales, the distance measure (or measures) must be chosen carefully. Several distance measures are employed for clustering [67, 68]. The most popular and commonly used one is the Euclidean distance,

$$d(X_i, X_j) = \sqrt{(X_i - X_j)'(X_i - X_j)} = \left[\sum_{l=1}^{p} (x_{il} - x_{jl})^2 \right]^{1/2} \tag{3}$$

which is the line of sight distance between two points representing the objects.

5.2 Genetic-algorithm-based clustering

Let us consider a set consisting of n vectors $\mathcal{X} = \{\mathbf{x}_1, \mathbf{x}_2, \dots, \mathbf{x}_n\}$, which will be clustered into k groups of homogeneous data. Each $\mathbf{x}_i \in R^d$ is a vector in a feature space of d real-valued measurements for defining the object features denoted by $\mathbf{x}_i$. The features can be color, breath, length, etc.

We have discussed data clustering with a GA-based algorithm. This algorithm is dependent on the splitting and merging techniques [69]. Initially the data are split into several sub-clusters, and in the next step those sub-clusters are merged to find out the required clusters. We have considered various types of datasets to show how the data are grouped to isolate properly the required number of clusters.

5.2.1 Cluster partitioning in R^2 feature space

We have studied seven datasets in R^2 feature space depicted in **Figures 4–10**. Among them, both **Figures 4** and **6** consist of three clusters, although the nature of clusters is different. In both figures, two clusters are located close to each other, whereas the third one is placed far from those two clusters. However, in **Figure 4**, the density of closely placed clusters is nonuniform. On the other hand, the third one is three times in size compared with other two clusters. The data points are more dense at the cluster center. The density gradually gets reduced toward the boundary. **Figure 4(a)** shows the original dataset, and **Figure 4(b)** illustrates how the clusters are isolated after using the genetically guided method.

The dataset in **Figure 5(a)** comprises five clusters. All of them are denser near the center and lighter toward the boundary. Three clusters are equal in size, and two of them are close to each other. The five clusters have been correctly isolated using the algorithm, as shown in **Figure 5(b)**.

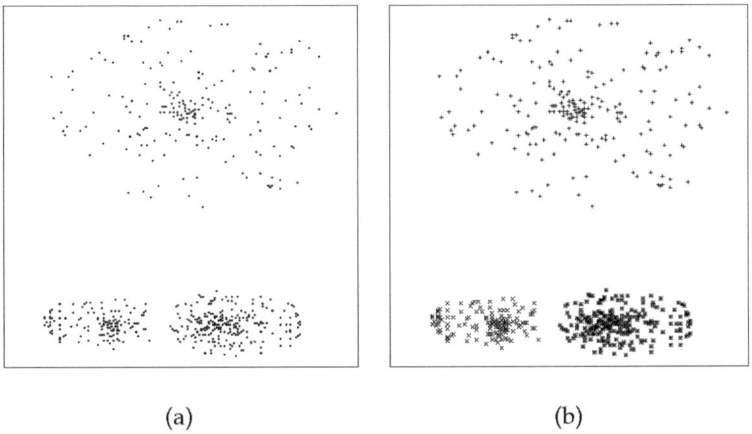

Figure 4.
(a) The original dataset with three clusters before the application of GA-based algorithm. (b) Isolated three clusters after completion of the algorithm [69].

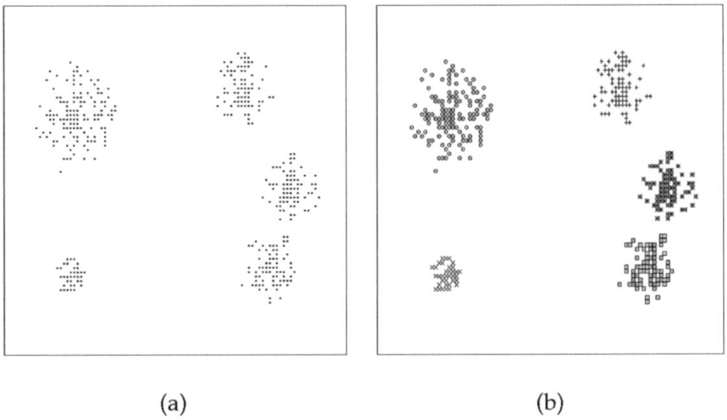

Figure 5.
(a) The original dataset with five clusters before the application of algorithm. (b) Isolated five clusters after the completion of algorithm [69].

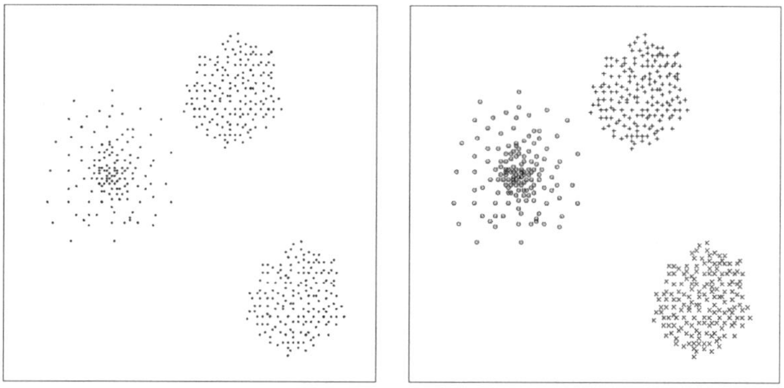

(a) (b)

Figure 6.
(a) The original dataset with three clusters before the application of algorithm. (b) Isolated three clusters after the completion of algorithm [69].

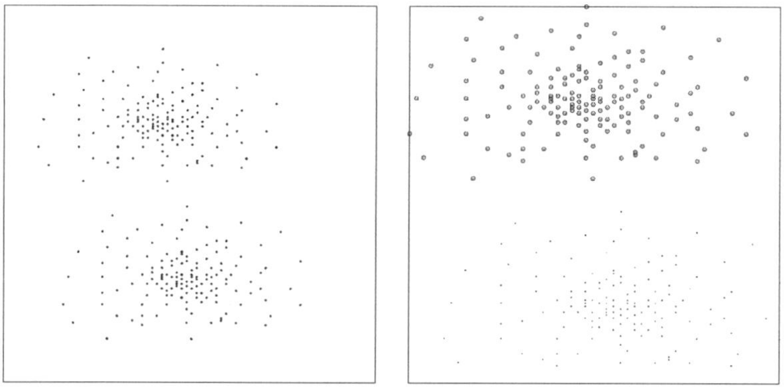

(a) (b)

Figure 7.
(a) The original dataset with two clusters before the application of algorithm. (b) Isolated two clusters after the completion of algorithm [69].

On the contrary, **Figure 6** consists of two closely located clusters of different densities. One cluster has uniformly dense data and the other has density tapering away from the center. The data density of the third cluster is also uniform. However, all of them are equal in size. **Figure 6(a)** and **(b)** illustrate the scenario, respectively, before and after the application of algorithm. All three clusters are properly isolated.

Figure 7 depicts two almost overlapping clusters of equal size and density. The data points distribution of both the clusters is interesting since they follow Gaussian distribution in R^2 space. The clusters are so closely placed that they seem to be visually a single cluster. However, the genetic algorithm can correctly identify the clusters, as illustrated in **Figure 7(b)**.

The dataset of **Figure 8** is different and interesting from previous four datasets. Here, one cluster is completely enclosed by the other. The data point density of both clusters is almost uniform. In the dataset, a special feature of the algorithm is used. It

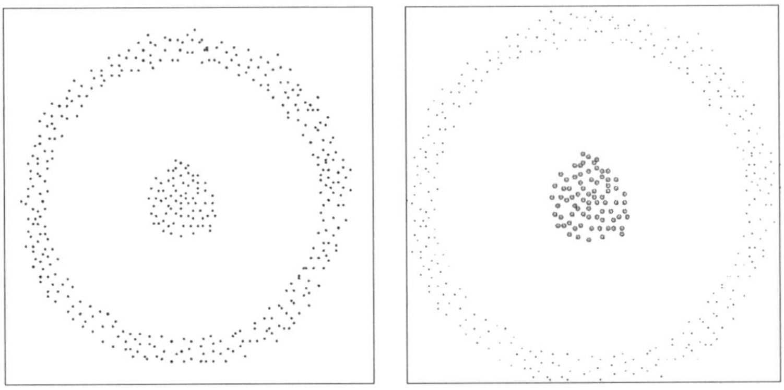

(a) (b)

Figure 8.
(a) The original dataset with two clusters before the application of algorithm. (b) Isolated two clusters after the completion of algorithm [69].

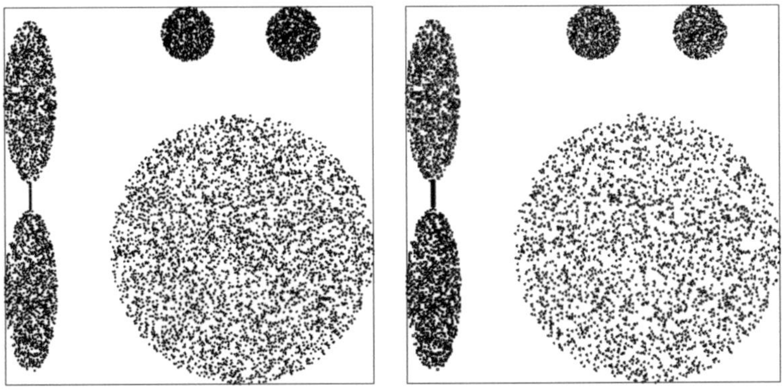

(a) (b)

Figure 9.
(a) The original dataset with six clusters before the application of algorithm. (b) Isolated six clusters after the completion of algorithm [69].

is to check the adjacency condition of sub-clusters, which are to be merged to isolate the clusters ultimately as depicted in 8(b).

Figures 9 and **10** consist of special type of clusters. **Figure 9(a)** is a combination of six clusters, which are different in density as well as size. Among six clusters, two are identical with elliptical shape and both are connected with a string of highly dense data points. They appear visually as a single cluster although they are disjoint. The set of clusters are shape-wise three in numbers and have been identified as three clusters. Here, the string of points is defined as a third cluster, and it connects those two elliptical shaped clusters. In **Figure 9(a),** two of the remaining three clusters are almost equal in shape and density. The largest cluster in **Figure 9(a)** has lesser density compared with other two clusters. However, all six clusters including the string of points are correctly isolated after application of the genetic algorithm as shown in **Figure 9(b).**

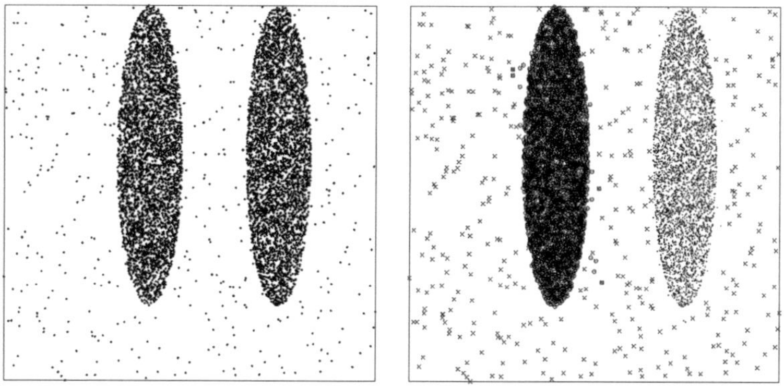

(a) (b)

Figure 10.
(a) The original dataset with two clusters and noise before the application of algorithm. (b) Isolated two genuine clusters and separated noise after the completion of algorithm [69].

The data shown in **Figure 10(a)** are quite different from all other datasets. It is also a special dataset in nature because it consists of two identical shaped clusters with different density in the presence of random noise scattered all over the 2-D feature space. Two clusters are perfectly identified and also the noise, which is identified as the third cluster as shown in **Figure 10(b)**. After application of the algorithm it is seen that the noise is clustered locally in smaller arbitrary shape in the space due to inherent characteristics of the random noise. The third cluster is then formed by merging all smaller clusters of noisy data as depicted in **Figure 10(b)**.

5.2.2 Cluster separation in R^n feature space

We have now discussed the cluster identification of a dataset of more than two features. Iris data is a set of four features [70], which is one of the most popular databases in the pattern recognition literature. The dataset contains three classes named as Iris Setosa (Class A), Iris Versicolor (Class B), and Iris Virginica (Class C). Each class has 50 instances and refers to a kind of Iris plant. The four feature attributes of the data are *Petal length*, *Petal width*, *Sepal length,* and *Sepal width*. Among the three classes, two are not linearly separable from each other, whereas the third one is identified separately from other two classes.

One cluster has been separated clearly from other two after invoking the algorithm. However, other two clusters are not perfectly separable (see **Table 4**).

	Classified as class A	Classified as class B	Classified as class C
Actual class A	50	0	0
Actual class B	0	40	10
Actual class C	0	0	50

Table 4.
Confusion matrix for Iris flower classification.

6. Shape recognition with genetic algorithm

The processing of dot patterns (DPs) or point sets in a plane is useful and important in pattern recognition problems. Dot patterns are encountered in various problems including points in feature space [71], pixels in digital image [72], physical objects such as stars in the galaxy [73], or spatial data [74–76].

An attribute of dot patterns is the visual shape generated by it. One simple way of defining the external shape is the convex hull [77]. But in most cases, the underlying shape from which the points emerge is not convex. To detect the nonconvex shape, the GA-based split and merge procedure [78] starts with the initial convex hull polygon. The *Splitting* process first divides the sides of the polygon. Consider a side $\overline{AB}$ and evaluate the average distance of r neighboring points of A and B. Now, if the average distance is smaller than the length of $\overline{AB}$, the side $\overline{AB}$ is considered as a candidate for splitting. Eventually such development takes care of the concavity of the DP by generating a zigzag polygonal border as depicted in **Figure 11(a)**. The *merging* algorithm is then applied for having a smoother border. Splitting results in inclusion of additional edges while merging does just the opposite, making the border looks less zigzag as in **Figure 11(b)**.

A DP may also be the union of more than one disjoint smaller DP region as shown in **Figure 11(c)**. The isolation process separates such regions by removing two lines of the resulting polygon simultaneously, which act as the bridge between two disjoint regions. The removal of lines is possible if the region within the lines does not contain any dot.

Now, consider a set S that contains n points in the plane and the convex hull consists of the points of S. In that case, the underlying shape of the pattern is the convex hull if the dot pattern does not contain any concavity, else the splitting of

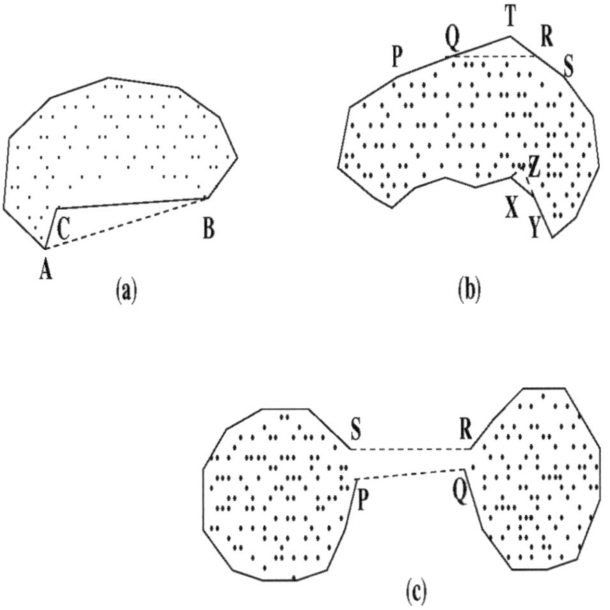

Figure 11.
(a) Edge (represented by dashed line) marked for splitting process (edge AB before splitting; edges AC and BC created by splitting). (b) Edge marked for merging process over region XYZ and region PQRS). (c) Edges (represented by dashed lines) marked for isolation process [78].

edges is required. **Figure 11** shows the splitting and merging of the edge/regions and the region for isolation.

6.1 Polygonal shape detection of dot pattern

The algorithm is studied on different dot pattern sets in R^2 space. The datasets are largely classified into three groups. Among these groups one consists of dot patterns with almost uniform density, as in **Figures 12** and **13**. **Figure 14** represents a dot pattern of variable density data points, and the data pooled from Gaussian distribution is shown in **Figure 15**. A special dot pattern with multiple distinct components is illustrated in **Figure 16**.

In the group of datasets with nearly uniform density data points, **Figures 12** and **13** show a DP with a concavity of spiral shape and a star-shaped DP, respectively. In **Figure 12**, the splitting process starts with the convex hull of 35 edges and ends with a

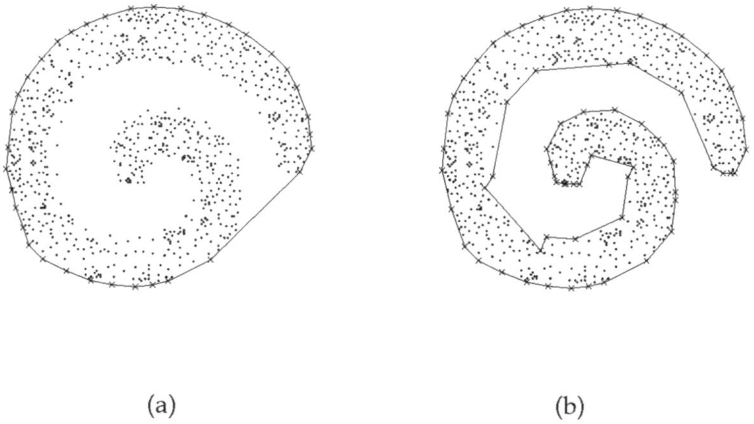

Figure 12.
(a) A spiral shaped DP and its convex hull. (b) Resulting perceptual border of the DP [78].

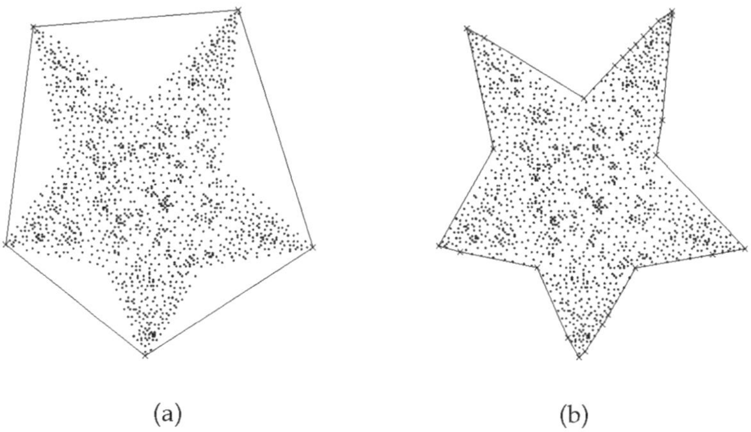

Figure 13.
(a) A star-shaped DP and its convex hull. (b) Resulting perceptual border of the DP with the GA-based merging process [78].

zigzag border polygon of 187 sides. After the splitting process, the merging algorithm is applied on the polygon since the DP does not contain any isolated component. Finally, a polygon of 68 edges is generated with a smooth border, as in **Figure 12(b)**. **Figure 13(a)** shows a polygon consisting of five concavities and the splitting process is executed on each side of the five-side convex hull. The resulting polygon of 181 edges is formed with a zigzag border. The GA-based procedure is applied on this polygon to produce a polygon (of 26 sides) with a star-like perceptual border as in **Figure 13(b)**.

The dot patterns of **Figures 14** and **15** are chosen to show the performance of the algorithm on variable density data points. Each DP consists of two concavities. In **Figure 14**, the upper half of the DP (the denser region) has one concavity and the other one is in the lower half of the DP (the lesser density region). The density of dots

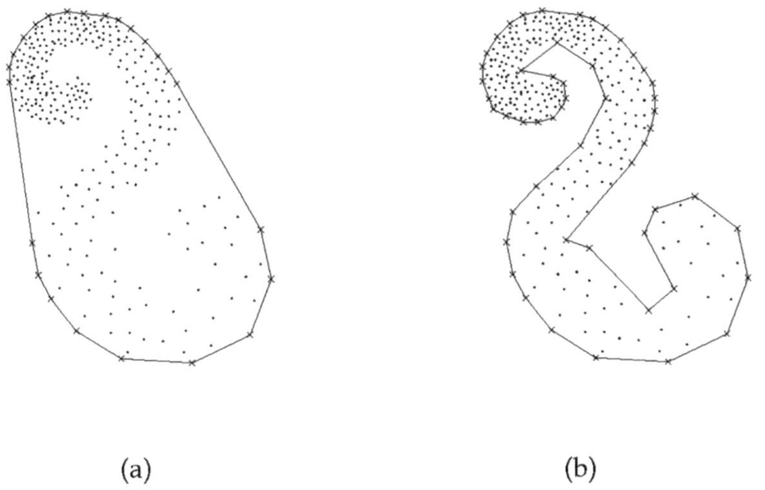

Figure 14.
(a) A DP with nonuniform density and its convex hull. (b) Resulting perceptual border of the DP [78].

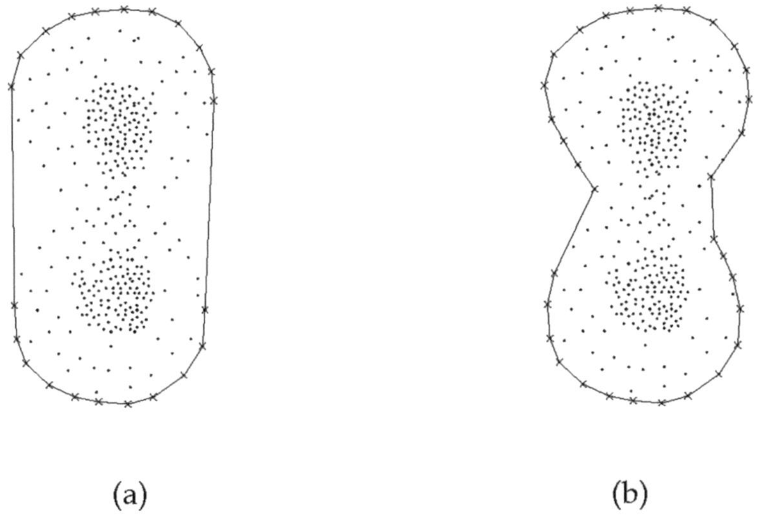

Figure 15.
(a) A DP with two Gaussian distributions and its convex hull. (b) Resulting perceptual border of the DP applying GA-based merging process [78].

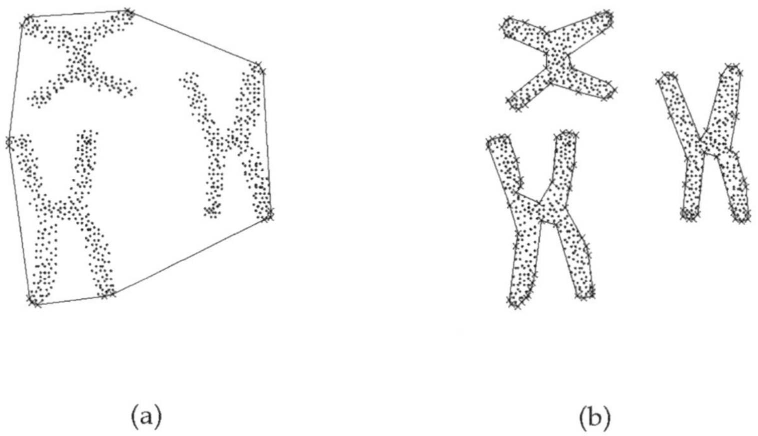

Figure 16.
(a) A multicomponent chromosome-like dot pattern and its convex hull. (b) Resulting perceptual border of the components [78].

in the dot pattern is gradually reduced toward the lower half. The DP of **Figure 15** also contains two concavities located in the middle region and opposite to each other. Its density decreases gradually from the center of each half toward the boundary portion. We have employed the GA-based merging technique on this dataset at the end of the splitting process. In **Figure 14**, the splitting procedure is started with a convex hull polygon of 24 sides (see **Figure 14(a)**) and a polygon of 102 sides is generated. Next, the merging process is executed over it and a 52-edge polygon with a smoother border is obtained as in **Figure 14(b)**. In **Figure 15(a)**, the splitting process starts on a 22-side convex hull polygon to create a polygon of 40 edges. The application of merging process finally generates a 32-side smother border polygon as shown in **Figure 15(b)**.

The ultimate polygons with smother border as in **Figure 17(b)** and **(c)** are generated from the starting polygon of **Figure 17(a)** if the user specifies the number of sides. **Figure 17(a)** shows a *C*-shape DP with its convex hull. In both cases, the splitting process is continued as long as there is no violation of the density-length condition. Now, the ultimate polygonal shape as in **Figure 17(b)** is obtained depending on the user's input of a 36-side polygon (say). Similarly, the polygon of **Figure 17(c)** is produced from **Figure 17(a)** when the user specifies 23 as the number of sides of polygon (say).

Figure 16 is a special dot pattern of three components where each one contains nearly uniformly dense data points. The convex hull of the DP is a 17-side polygon. The splitting process followed by the isolation process separates the DP into three distinct regions with zigzag polygonal boundaries. Finally, for each isolated polygon, the merging process generates three separated polygons with a smoother border as in **Figure 16(b)**.

7. Biological sequence alignment with genetic algorithm

An alignment at sequence level represents the primary structure of biological macromolecules, which is represented as a linear sequential chain of residues. Residues are the building blocks of macromolecules in which DNAs and RNAs are made of nucleotide residues and proteins are of amino acid residues. The building blocks encode the

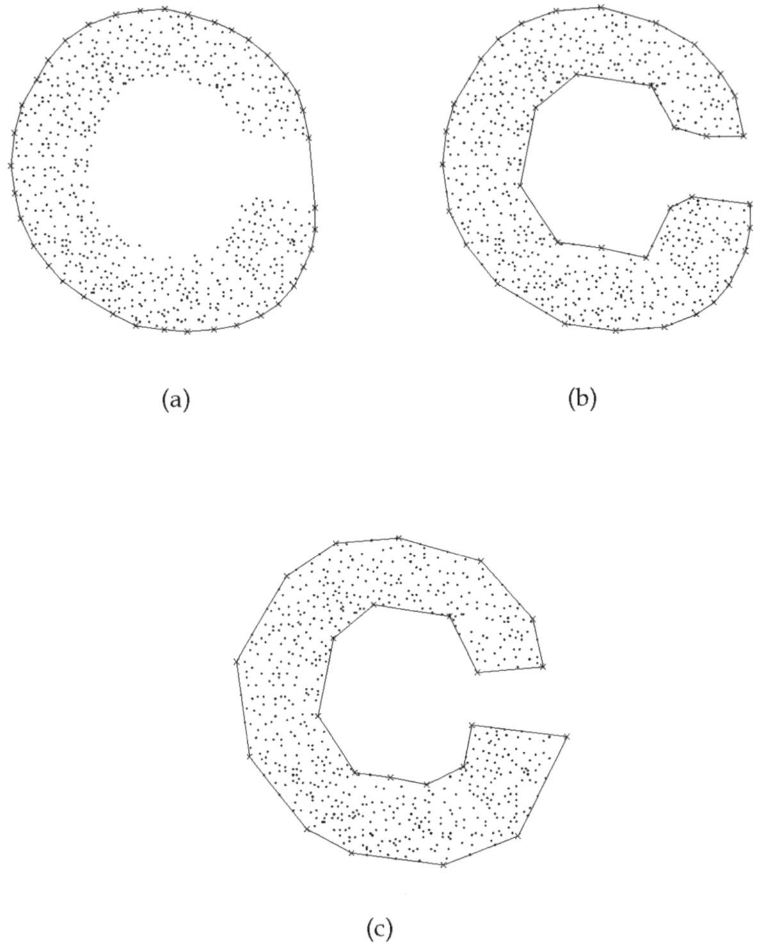

Figure 17.
(a) A C-shaped DP and its convex hull. (b) Perceptual border of the DP with 36 edges (user specified). (c) Perceptual border of the DP with 23 edges (user-specified) [78].

history of molecular evolution. During the process of evolution, residues accumulate many random changes and diverge over time. Some of the changes are accepted and selected by natural selections, and some of them are not. Residues, which are involved in important functional and/or structural roles, tend to be preserved by natural selection and do not accumulate random changes [79]. These parts of sequences help to infer the evolutionary history and identify the common ancestor of two or more related sequences. Therefore, the sequences that share the common evolutionary origin are called homologous sequences. Homology is the conclusion that can be drawn by looking at the degree of similarities between two or more sequences of different species.

7.1 Scoring schemes in sequence alignment

The similarity between two or multiple sequences is quantified by measuring the alignment score. To choose the best alignment among a set of sequences, quantification of alignment is required. It identifies the evolutionary distances between two or multiple sequences. The best alignment always associated with the maximum shared similarity or identity and the lowest number of mutational events between the sequences. Therefore, the scoring systems always score positive when there are identical or similar residues in the alignment.

On the other hand, the mutational events are always associated with zero or negative scores. Different scoring schemes are used in sequence alignment to know the level of identity or similarity. The process of score evaluation for a pairwise sequence alignment is more straightforward and simple compared with scoring the alignment of multiple sequences together. In two aligned sequences, the total number of identical residues yields a percentage identity between them. This identity count is used mostly for nucleotide sequence alignment, as the nucleotides A, T/U, G, and C play equivalent roles in the structure and function of the DNA or RNA molecule. Therefore, the nucleotides are either identical, which is a match in alignment, or nonidentical, which is a mismatch in an alignment. In contrast, for protein sequences, a similarity score is calculated along with the identity score denoting the amino acids having similar physicochemical properties. The substitution matrices are normally consulted for protein sequence alignment.

7.2 Sequence alignment scheme without gaps

The GA-based sequence alignment [80] is a simple method to align a pair of sequences (a query sequence and a known database sequence) without gaps for finding similarity. When a biological sequence (DNA/RNA/protein), called the query sequence is given, one usually performs a similarity measure within the databases that consist of all available genomic sequences and the known protein sequences. Eventually, the search yields many sequences with varying degree of similarities. It then depends on the user to identify those that are associated with the higher scores and homologous. In the alignment technique, we consider one sequence as a database sequence denoted by D and the other as a query sequence denoted by Q.

In the initialization step, an initial population of size N is randomly generated as the probable alignment solutions for the input Q. Each individual $\mathbf{P_i}, \forall i \in \{1, 2, \dots, N\}$ or a chromosome is a string of bits such that $\mathbf{P_i} = b_1 b_2 \dots b_L$ where L is the length of each individual/chromosome, which is equal to the given Q, and each b_j is $\{0, 1\}, \forall j \in \{1, 2, \dots, L\}$. The *1*'s in the chromosome represent the presence of the residues in the corresponding positions and the *0*'s indicate the absence of the residues in Q.

In this approach, all three operators of the conventional GA, namely tournament selection, one-point crossover, and bit-flip mutation are used [1]. In each iteration, the algorithm advances the search toward a better solution by producing a better population. After random selection of two chromosomes from the population pool, the crossover and the mutation operations are performed on them.

The fitness score of a chromosome is evaluated by the following fitness function F.

$$W = \sum_{i=1}^{n} W_i \tag{4}$$

where n is the total number of residues to be aligned, and w is the weight of the alignment score.

For protein sequence alignment, the BLOSUM62 matrix is consulted and for DNA sequence, an identical match gets +1 score, and a mismatch gets 0. The fitness score determines the similarity or identity level between two sequences D and Q. The fitness score is now calculated for pairwise alignment between two sequences D and Q. It is evaluated in three ways. The scoring is for the straight alignment and the alignments with a left or right shift.

For a straight alignment, let us consider the following two nucleotide sequences, *D* and *Q*.

$$D = A\ T\ G\ C\ T\ T\ A\ G\ T\ C$$
$$Q = A\ C\ G\ C\ A\ T\ A\ G\ A\ C$$

Let us now consider the following binary coded chromosome from the population of an intermediate generation as a probable solution of query sequence Q_p.

$$Q_p = 0\,1\,1\,0\,1\,0\,1\,1\,0\,1$$

The equivalent decoded value of Q_p by replacing 1 with the corresponding residue of *Q* will be as follows.

$$Q_p = \_C\,G\_A\_A\,G\_C$$

where "_" denotes the presence of 0 in Q_p.

Therefore, the alignment score or the fitness of the chromosome would be 4 and the alignment structure looks like:

$$D = A\ T\ G\ C\ T\ T\ A\ G\ T\ C$$
$$Q_p = \_C\,G\_A\_A\,G\_C$$

However, the alignment by shifting is necessary for a different representation of Q to achieve the best alignment score. For example, if Q is ACTTAGTAAC, then the shifting to the right is required to obtain the optimum alignment with alignment score 6, which is illustrated below,

$$D = A\ T\ G\ C\ T\ T\ A\ G\ T\ C$$
$$Q = A\ C\ T\ T\ A\ G\ T\ A\ A\ C$$

Similarly, for Q: GCGTTGCTTAGA, the left shift alignment is necessary to obtain the best alignment with score 7, which is shown below,

$$D = A\ T\ G\ C\ T\ T\ A\ G\ T\ C$$
$$Q = G\ C\ G\ T\ T\ G\ C\ T\ T\ A\ G\ A$$

For the evaluation of the final fitness score of an individual, the scores of all three alignments (straight, shift right, and shift left) techniques are considered. However, the selection of position to start alignment depends on the randomly chosen position number. Since minimum 30% matching between two sequences is necessary to evaluate the homologous relationship [81, 82], the random numbers between 1 and 70% of the length of the sequence is generated. For example, if a sequence length is of 100 residues, we shall identify a position randomly between 1st and 70th residues of D or Q depending on the alignment type. The length of D is considered for the straight as well as the right shift alignments. The length of D is also considered for the left shift alignment if it is smaller than or equal to Q. However, the length of Q is taken when D is larger than Q. The total number of times for choosing the position numbers for these three types of alignments depends on the average length of D and Q. In this case, it is 30% of the average length. For example, if the average length of D and Q is 200, the alignment techniques will be continued 60 times with different random starting position values. Now the best matching score is extracted from 60 results given by three alignment techniques.

8. Conclusions

This chapter describes the application of genetic algorithm in various scientific fields as a reliable optimizer. It is narrated how the simple/conventional genetic algorithm has been advanced with time. Apart from these some real applications of the modified genetic algorithm with illustrations have been elaborated in brief (references are provided for detailed description) on the synthetic data. It is shown how the GA identifies a pattern (in single/multiple dot(s) or edge map(s) scene) in 2-D space. The reasoning is given to explain the differences in results of matching scores of two genetic sequential and distributed schemes (CGA and CAGA or DCGA and DCAGA). A genetic-based clustering approach with results provides the efficient use of GA for grouping synthetic as well as real data. The function of GA in complex shape detection and biological sequence alignment on synthetic data further shows the application of GA as a useful and essential optimizer.

Abbreviations

GA	Genetic Algorithm
SVM	Support Vector Machine
CGA	Conventional Genetic Algorithm
CAGA	CAscaded Genetic Algorithm
DCAGA	Distributed CAscaded Genetic Algorithm
2-D	Two-Dimensional
3-D	Three-Dimensional
R^2	Two-Dimensional space
DP	Dot Pattern
DNA	Deoxyribonucleic acid
RNA	Ribonucleic acid

Author details

Gautam Garai
Saha Institute of Nuclear Physics, Kolkata, India

*Address all correspondence to: gautam.garai.16@gmail.com

References

[1] Goldberg DE. Genetic Algorithms in Search, Optimization and Machine Learning. Reading, MA: Addison-Wesley Publishing; 1989

[2] Rechenberg I. Cybernetic solution path of an experimental problem. Royal Aircraft Establishment (U.K.): Ministry of Aviation; 1965

[3] Schwefel HP. Evolutions Strategie and Numerische Optimierung. Ph.D. Thesis. Berlin: Technische University; 1975

[4] Fogel LJ, Owens AJ, Walsh MJ. Artificial Intelligence Through Simulated Evolution. New York: John Wiley; 1966

[5] Box GEP. Evolutionary operation: A method for increasing industrial productivity. Journal of the Royal Statistical Society, Vol. C. 1957;**6**(2): 81-101

[6] Friedman GJ. Digital simulation of an evolutionary process. General Systems Yearbook. 1959;**4**:171-184

[7] Bremermann HJ. Optimization through evolution and recombination. In: Yovits MC, Jacobi GT, Goldstein GD, editors. Self-Organizing Systems. Washington D. C.: Spartan Books; 1962

[8] Reed J, Toombs R, Barricelli NA. Simulation of biological evolution and machine learning. Journal of Theoretical Biology. 1967;**17**:319-342

[9] Baricelli NA. Numerical testing of evolution theories. Acta Biotheoretica. 1962;**16**:69-126

[10] Fraser AS. Simulation of genetic systems by automatic digital computers: I introduction. Australian Journal of Biological Sciences. 1957;**10**:484-491

[11] Martin GG, Cockerham CC. High speed selection studies. In: Kempthorne O, editor. Biometrical Genetics. USA: Pergamon; 1960

[12] Hill A, Taylor CJ. Model-based image interpretation using genetic algorithm. Image and Vision Computing. 1992;**10**: 295-300

[13] Roth G, Levine MD. Geometric primitive extraction using a genetic algorithm. IEEE Transactions on Pattern Analysis and Machine Intelligence. 1994; **16**(9):901-905

[14] Toet A, Hajema WP. Genetic contour matching. Pattern Recognition Letters. 1995;**16**:849-856

[15] Lin DS, Leou JJ. A genetic algorithm approach to Chinese handwriting normalization. IEEE Trans. Systems, Man and Cybernetics-Part B. 1997;**27**(6): 999-1007

[16] Yamany SM, Khiani KJ, Farag AA. Application of neural networks and genetic algorithms in the classification of endothelial cells. Pattern Recognition Letters. 1997;**18**:1205-1210

[17] Giacinto G, Paolucci P, Roli F. Application of neural networks and statistical pattern recognition algorithms to earthquake risk evaluation. Pattern Recognition Letters. 1997;**18**:1353-1362

[18] Wang YK, Fan KC, Horng JT. Genetic-based search for error-correcting graph isomorphism. IEEE Transactions on Systems, Man and Cybernetics-Part B. 1997b;**27**(4):588-596

[19] Ansari N, Chen MH and Hou ESH. Point pattern matching by a Genetic Algorithm. In: Proc. of the 16th Annual

Conf. of IEEE Industrial Electronic Society (IECON'90), Vol. II. Pacific Grove: 1990. pp. 1233-1238

[20] Mirmehdi M, Palmer PL, Kittler J. Genetic optimization of the image feature extraction process. Pattern Recognition Letters. 1997;**18**:355-365

[21] Chan KCC, Lee V, Leung H. Generating fuzzy rules for target tracking using a steady-state genetic algorithm. IEEE Transactions on Evolutionary Computing. 1997;**1**(3):189-200

[22] Schnecke V, Vornberger O. Hybrid genetic algorithms for constrained placement problems. IEEE Transactions Evolutionary Computing. 1997;**1**(4): 266-277

[23] Zhang F, Zhang YF, Nee AYC. Using genetic algorithms in process planning for job shop machining. IEEE Transactions on Evolutionary Computing. 1997;**1**(4):278-289

[24] Chen S, Wu Y, McLanghlin S. Genetic algorithm optimization for blind channel identification with higher order cumulant fitting. IEEE Transactions on Evolutionary Computing. 1997;**1**(4): 259-265

[25] Kushchu I. Web-based evolutionary and adaptive information retrieval. IEEE Transactions on Evolutionary Computation. 2005;**9**(2):117-125

[26] Raidl GR, Koller G, Julstrom BA. Biased mutation operators for subgraph-selection problem. IEEE Transactions on Evolutionary Computation. 2006;**10**(2): 145-156

[27] Cavicchio DJ. Adaptive Search using Simulated Evolution. Ph.D. dissertation. Ann Arbor, Michigan: University of Michigan; 1970

[28] De Jong KA. An Analysis of Behavior of a Class of Genetic Adaptive System. Doctoral dissertation. Michigan: University of Michigan; 1975

[29] Fogel DB. Evolutionary Computation: Toward a New Philosophy of Machine Intelligence. Piscataway, NJ: IEEE Press; 1995

[30] Back T, Schwefel HP. An overview of evolutionary algorithm for parameter optimization. Evolutionary Computation. 1993;**1**:1-23

[31] Davis LD. Handbook of Genetic Algorithms. New York: Van Nostrand Reinhold; 1991

[32] Harp SA, Samad T. Genetic synthesis of neural network architecture. In: Davis L, editor. Handbook of Genetic Algorithms. New York: University of Chicago Press; 1992. pp. 202-221

[33] Rizzi S. Genetic operators for hierarchical graph clustering. Pattern Recognition Letters. 1998;**19**:1293-1300

[34] Kim D, Ahu S. A MS-GS VQ codebook design for wireless image communication using genetic algorithms. IEEE Transactions on Evolutionary Computation. 1999;**3**(1):35-52

[35] Maniezzo V. Genetic evolution of the topology and weight distribution of neural networks. IEEE Transactions on Neural Networks. 1994;**5**:39-53

[36] Cantú-Paz E. A summary of research on parallel genetic algorithms. Illinois Genetic Algorithm Lab., Univ. Urbana, IL: Illinois Genetic Algorithm Lab., Univ. Illinois Urbana-Champaign; 1995. p. 950076. Tech. Rep

[37] Tomassini M. Parallel and distributed evolutionary algorithms: A review. In: Miettinen K, Mkel M,

Neittaanmki P, Periaux J, editors. Evolutionary Algorithms in Engineering and Computer Science. New York: Wiley; 1999. pp. 113-133

[38] Veldhuize DAV, Zydallis JB, Lamont GB. Considerations in engineering parallel multiobjective evolutionary algorithms. IEEE Transactions on Evolutionary Computation. 2003;7:144-173

[39] Hao JK, Dome R. A new population-based method for satisfiability problems. In: Proc. of the 11th European Conference on Artificial Intelligence. New York: Wiley; 1994. pp. 135-139

[40] Holland JH. Adaptation in Natural and Artificial Systems. Ann Arbor, MI: Univ Michigan Press; 1975

[41] Lienig J. A parallel genetic algorithm for performance-driven VLSI routing. IEEE Transactions on Evolutionary Computation. 1997;**1**:29-39

[42] Sena GA, Megherlu D, Isern G. Implementation of a parallel genetic algorithm on a cluster of workstations: Traveling salesman problem, a case study. Future Generation Computer Systems. 2001;**17**:477-488

[43] Easton FF, Mansour N. A distributed genetic algorithm for deterministic and stochastic labor scheduling problems. European Journal of Operational Research. 1999;**118**:505-523

[44] Notredame C, Higgins DG. SAGA: Sequence alignment by genetic algorithm. Nucleic Acids Research. 1996; **24**(8):1515-1524

[45] Arenas DE, Ochoterena H, Rodriguez VK. Multiple sequence alignment using a genetic algorithm and GLOCSA. Journal of Artificial Evolution and Applications. 2009;**2009**: 963150

[46] Naznin F, Sarker R, Essam D. Vertical decomposition with genetic algorithm for multiple sequence alignment. BMC Bioinformatics. 2011;**12**: 353

[47] Shyu C, Foster J. Evolving Consensus Sequence for Multiple Sequence Alignment with a Genetic Algorithm. In: Cantú-Paz E et al., editors. Proc. Conf. Genet. and Evol. Comp. (GECCO'03), vol. 2724. Berlin Heidelberg, Chicago, IL, USA: LNCS, Springer-Verlag; 2003. pp. 2313-2324

[48] Michalewicz Z, Fogel DB. How to solve it: modern heuristics. 2nd rev. and extended. ed. Berlin; London: Springer; 2004

[49] Narimani Z, Beigy H, Abolhassani H. A new genetic algorithm for multiple sequence alignment. International Journal of Computational Intelligence and Applications. 2012;**11**(04):1250023

[50] Zhao S, Jin R, Abroshan H, Zeng C, Zhang H, House SD. Gold nanoclusters promote electrocatalytic water oxidation at the nanocluster/cose2 interface. Journal of the American Chemical Society. 2017;**139**:1077-1080. DOI: 10.1021/jacs.6b12529

[51] Bader SD. Colloquium: Opportunities in nanomagnetism. Reviews of Modern Physics. 2006;**78**: 1-15. DOI: 10.1103/RevModPhys.78.1

[52] Pelegrini M, Parreira RLT, Ferrão LFA, Caramori GF, Ortolan AO, Silva EH. Hydrazine decomposition on a small platinum cluster: The role of n2h5 intermediate. Theoretical Chemistry Accounts. 2016;**135**:58. DOI: 10.1007/ s00214-016-1816-x

[53] Islas R, Heine T, Ito K, Schleyer P, v. R., and Merino G. Boron rings enclosing planar hypercoordinate group 14 elements. Journal of the American Chemical Society. 2007; **129**:14767-14774. DOI: 10.1021/ja07 4956m

[54] Sathya M, Jayaselvi M, Joshi S, Pandy E, Pareek PK, Jamal SS, et al. Cancer categorization using genetic algorithm to identify biomaker genes. Journal of Healthcare Engineering. 2022; **ID**:5821938

[55] Xu Y, Zeng M, Liu Q, Wang X. A genetic algorithm based multilevel association rules mining for big datasets. Mathematical Problems in Engineering. 2014;**2014**:867149

[56] Devaraj N. Feature Selection using Genetic Algorithm to Improve SVM Classifier. USA: LAP LAMBERT Academic Publishing; 2019

[57] Bhasin H, Bhatia S. Application of genetic algorithms in machine learning. Intl. Journal of Computer Science and Information Technologies. 2011;**2**(5): 2412-2415 0975-9646

[58] Manszoori TK, Suman A, Mishra AK. Application of genetic algorithm for cancer diagnosis by feature selection. Intl. Journal of Engineering Research and Technology. 2014;**3**(8):1295-1301 2278-0181

[59] Resmini R, Silva L, Aranjo AS, Medeiros P, Muchaluat-Saade D, and Conci A. Combining genetic algorithms and SVM for breast cancer diagnosis using infrared thermography. Sensors. 2021;**21**:4802

[60] Griffin PM, Alexopoulos C. Point pattern matching using centroid bounding. IEEE Transactions on Systems, Man, and Cybernetics. 1989; **19**(5):1274-1276

[61] Lavine D, Lambird BA, Kanal LN. Recognition of spatial point patterns. Pattern Recognition. 1983;**16**(3):289-295

[62] Sprinzak J, Werman M. Affine point matching. Pattern Recognition Letters. 1994;**15**(4):337-339

[63] Zhang L, Xu W. Point-pattern matching using irreducible matrix and relative invariant. Tsinghua Science and Technology. 1999;**4**(4): 1602-1605

[64] Garai G, Chaudhuri BB. A cascaded genetic algorithm for efficient optimization and pattern matching. Image and Vision Computing. 2002;**20**: 265-277

[65] Garai G, Chaudhuri BB. A distributed hierarchical genetic algorithm for efficient optimization and pattern matching. Pattern Recognition. 2007;**40**:212-228

[66] Jain AK. Fundamentals of Digital Image Processing. Englewood Cliffs, N.J: Prentice-Hall; 1989

[67] Everitt BS. Cluster Analysis. London: Edward Arnold; 1993

[68] Johnson RA, Wichern DW. Applied Multivariate Statistical Analysis. New Jersey: Prentice Hall; 1992

[69] Garai G, Chaudhuri BB. A novel genetic algorithm for automatic clustering. Pattern Recognition Letters. 2003;**25**:173-187

[70] Newman DJ, Hettich S, Blake CL and Merz CJ. UCI Repository of machine learning databases, 1998. Univ. of California, Irvine: Dept. of Information and Computer Sciences; 1998. Available from: http://www.ics.uci.edu/~mlearn/ MLRepository.html

[71] Duda RO, Hart PE. Pattern Classification and Scene Analysis. New York: John Wiley & Sons, Inc.; 1973

[72] Faugeras O. Three-Dimensional Computer Vision: A geometric Viewpoint. Cambridge: The MIT Press; 1993

[73] Ogawa H. Labeled pattern matching by Delaunay triangulation and maximal cliques. Pattern Recognition. 1986;**19**: 35-40

[74] Taylor PJ. Quantitative Methods in Geography: An Introduction to Spatial Analysis. Boston: Houghton Mifflin Company; 1977

[75] Laurini R. and Thompson D. Fundamental of Spatial Information Systems. The A.P.I.C. Series, No. 37. London: Academic Press; 1992

[76] Okabe A, Boots B, Sugihara K. Spatial tessellations: Concepts and Applications of Voronoi Diagrams. New Jersey: John Wiley and Sons; 1992

[77] Ronse C. A bibliography on digital and computational convexity (1961-1988). IEEE Transactions on Pattern Analysis and Machine Intelligence. 1989;**11**:181-190

[78] Garai G, Chaudhuri BB. A split and merge procedure for polygonal border detection of dot pattern. Image and Vision Computing. 1999;**17**:75-82

[79] Xiong J. Essential bioinformatics. NY: Cambridge University Press; 2006

[80] Garai G, Chowdhury B. A novel genetic approach for optimized biological sequence alignment. Journal of Biophysical Chemistry. 2012;**3**:201-205

[81] Sander C, Schneider R. Database of homology-derived protein structures and the structural meaning of sequence alignment. Proteins. 1991;**9**(1):56-68

[82] Rost B. Twilight zone of protein sequence alignments. Protein Engineering. 1999;**12**(2):85-94

Section 2

Engineering Applications

Chapter 3

Power System Small-Signal Stability Enhancement Using Damping Controllers Designed Based on Evolutionary Algorithms

Komla Agbenyo Folly, Severus Panduleni Sheetekela and Tshina Fa Mulumba

Abstract

This chapter is concerned with the stability enhancement of a power system using power system stabilizers (PSSs) designed based on four evolutionary algorithms (EAs), namely, genetic algorithms (GAs), breeder genetic algorithm (BGA), population-based incremental learning (PBIL), and differential evolution (DE). GAs have been widely applied in many fields of engineering and science and have shown to be a robust and powerful adaptive search algorithm. However, GAs are known to have several limitations. To deal with these limitations, many variant forms of GAs have been suggested often tailored to specific problems. In this research, we investigated the performances of GA-PSS and three other EAs-based PSSs (i.e., BGA-PSS and PBIL-PSS and DE-PSS) in improving the small-signal stability of a power system. These EAs have been selected on the basis of their simplicity, efficiency, and effectiveness in solving the optimization problem at hand. Frequency domain and time-domain simulation results show that DE-PSS, PBIL-PSS, and BGA-PSS performed better than GA-PSS. Time domain simulations suggest that overall, DE-PSS performs better than PBIL-PSS and BGA-PSS in terms of undershoot and subsequent swings, albeit with a relatively large first swing overshoot. The performances of BGA-PSS and PBIL-PSS are similar. On the other hand, GA-PSS gives a better response than the conventional PSS (CPSS).

Keywords: breeder genetic algorithm, damping ratio, genetic algorithms, differential evolution, low-frequency oscillations, power-system stabilizer, population-based incremental learning

1. Introduction

Over the past decades, low-frequency oscillatory modes have been a major concern to power system engineers [1]. These oscillatory modes ranging from 0.1 to 3 Hz tend to be poorly damped especially in moderately to heavily loaded systems that are

 IntechOpen

equipped with high gain, fast-acting automatic voltage regulators (AVRs) [2, 3]. Generally, we distinguish two main oscillation modes: local and inter-area modes. Local modes (0.8–2 Hz) involve local generators oscillating against each other. On the other hand, inter-area modes are caused by groups of generators in one part of the system swinging against other groups in the interconnected power system having frequencies ranging from 0.1 Hz to 0.8 Hz. Compared to local modes, inter-area modes are generally the most critical modes that need to be damped [4, 5]. These modes are found in almost all interconnected power stems. If they are not adequately damped, the oscillations may sustain and grow, and this may lead to system blackout. Power system stabilizers (PSSs) have been proposed to modulate low-frequency oscillations and increase the damping of electromechanical modes [1, 2]. Tuning the PSS parameters is not a trivial task. Power utilities have preferred using conventional PSSs (CPSSs) designed around a nominal operating condition. The design of the CPSS is generally based on conventional control approaches such as root locus, phase compensation, and pole placement techniques [1–5]. However, since these approaches are not robust, the designed CPSS tends to deviate from optimal operation when the system experiences a range of changes away from the nominal operating conditions. Therefore, new design approaches are required to design a PSS that can operate optimally under a wide range of operating conditions [3, 6]. Evolutionary algorithms (EAs) such as genetic algorithms (GAs) [7–12], differential evolution (DE) and its variants [13, 14], particle swarm optimization (PSO) [15], population-based incremental learning (PBIL) [16–19], and breeder genetic algorithms (BGA) [11, 20–24] are efficient heuristic search methods that are capable of solving complex optimization problems. They do not require the objective function to have properties such as continuity, smoothness, and differentiability. They have many advantages over traditional optimization methods and have attracted considerable attention in recent years. Many of these methods have been applied to power system damping controller design with encouraging results. In particular, GAs have been extensively used to solve global optimization problems in academia and are now being accepted by some industries [9]. DE, PBIL, and BGA are easy to implement yet efficient and robust in solving optimization problems. Therefore, they are considered in this work.

GAs are biologically motivated adaptive systems based on natural selection and genetics. GAs are generally used to solve optimization problems by the exploitation of a random search [7, 8]. Although GAs are seen to be robust and powerful adaptive search mechanisms, they have several drawbacks [9]. One of these drawbacks is related to "genetic drift." This phenomenon prevents GAs from maintaining diversity in their population. Other issues include the nonexistence of theoretical guidance for selecting optimal GA parameters such as population size, crossover, and mutation rates. Moreover, the natural selection approach used by GAs is not immune from failure [22]. Breeder genetic algorithm (BGA) has been proposed to cope with some of these drawbacks. It applies almost the same ideas as in GA, except that it is based on artificial selection as practiced in animal breeding rather than using natural selection based on Darwinian evolution [23, 24]. Artificial selection (selective breeding) refers to the intentional breeding for certain qualities or a combination of qualities [23]. This is in contrast with the natural selection that is the process whereby organisms survive and produce offspring by naturally adapting to their environment. Generally, individuals in BGA are represented as real numbers instead of binary or integers. The main advantage of using BGA over GA is its simplicity in the selection method and the fewer parameters. The major limitation of this algorithm is that there is a likelihood of premature convergence that could lead BGA to converge to the local optimum rather than the global one. To deal with the problem of premature convergence, an adaptive mutation is used [23, 24]. In this case, the mutation rate is not fixed but varies according to the convergence and performance of the population. This is the type of BGA that will be discussed later in this chapter.

Population-based incremental learning (PBIL) is a combination of GA and competitive learning. It extends the features of the evolutionary genetic algorithm (EGA) through the reexamination of the performance of the GA in terms of competitive learning [16–19]. It was originally proposed by Baluja [18, 19]. In PBIL, the crossover operator is removed, and the role of the population is redefined. PBIL works on probabilistic vectors (PVs), which control the random bit strings generated by PBIL. The PVs are used to create other vectors through competitive learning. The PV is then updated to increase the likelihood of producing solutions corresponding to the current best individual. It has been shown that PBIL is simpler than GA and in many cases performs better than GA and has less overhead [11, 16–19].

Differential evolution (DE) is a powerful stochastic optimizer whose search mechanism involves a differential mutation technique [12, 13, 25]. The algorithm is both simple and robust, with several variants exhibiting different tradeoffs between convergence speed and robustness. Most often DE outperforms its counterparts in efficiency and robustness [12–14, 25].

This chapter discusses the optimal design of power system stabilizers (PSSs) using four evolutionary algorithm (EAs) techniques, namely, genetic algorithms (GAs), breeder genetic algorithm (BGA) with adaptive mutation, population-based incremental learning (PBIL), and differential evolution (DE). For comparison purposes, the conventional PSS (CPSS) is also included in this work. The performance and effectiveness of the PSSs in damping the electromechanical modes are investigated using both frequency-domain analysis and time-domain simulations. Simulation results show that all the EA-based PSSs (GA-PSS, BGA-PSS, PBIL-PSS, and DE-PSS) perform better than the CPSS for all the operating conditions considered. Frequency domain simulation suggests that DE-PSS, PBIL-PSS, and BGA-PSS have similar performances in terms of the damping ratios that they provided. Time-domain simulations however suggest that overall, DE-PSS performs slightly better than PBIL-PSS and BGA-PSS in terms of undershoot and subsequent swings, albeit with a slightly large 1st swing overshoot. GA-PSS is shown to give the worst performance amount to the EAs. The chapter is organized as follows: Sections 2–4 present the overview of BGA, PBIL, and DE, respectively; Section 5 discusses the system model; Section 6 is concerned with the objective function; Section 7 presents the design of the PSSs; Section 8 discusses the simulation results; and the conclusions are presented in Section 9.

2. Overview of breeder genetic algorithm

As discussed previously, breeder genetic algorithm (BGA) is similar to genetic algorithms (GAs), with the exception that it uses artificial selection and has fewer genetic parameters. Also, BGA uses real-valued representation as opposed to GAs that mainly use binary and sometimes floating or integer representation. BGA is a versatile and effective function optimizer. It has the advantage of being simpler than GA. To deal with the issue of premature convergence that is common with BGA, a modified version of BGA called adaptive mutation BGA is used in this work [11, 20, 23]. In the truncation selection method that has been adopted, the T% of the fittest individuals is selected from the current population of N individuals and goes through recombination and mutation to form the next generation. The rest of the individuals are discarded. In the truncation method, the fittest individual in the population is automatically part of the next generation. The other top T%-1 goes through recombination and mutation to form the rest of the individuals in the next generation. The process is repeated until an optimal solution is obtained or the maximum number of iterations has been reached.

2.1 Recombination

Recombination is similar to a crossover in GAs. The adaptive mutation BGA proposed in this work allows various possible recombination methods to be used, each of them searching the space with a particular bias. Because we do not have prior knowledge as to which bias is likely to suit the optimization task, it is better to include several recombination methods and allow selection to do the elimination. Two recombination methods were used in this work: volume and line recombination [11].

In volume recombination, a random vector r_i equal to the parents' length is generated and the child c_i is produced by the following expression:

$$c_i = r_i a_i + (1 - r_i) b_i \tag{1}$$

Where c_i is a component of the child, a_i and b_i are the two respective parent components, and r_i is a random vector component.

The child can be said to be located at a point inside the hyper box defined by the parents as shown in **Figure 1**.

In line recombination, a single uniformly random number r is generated between 0 and 1, and the child is obtained as shown below [23].

$$c_i = r a_i + (1 - r) b_i \tag{2}$$

Where c_i, a_i, and b_i are defined as in Eq. (1).

2.2 Adaptive mutation

As mentioned before, one of the main concerns in GA has been the issue of premature convergence. This issue is also encountered in the classical BGA. This problem can be reduced in BGA by using an adaptive mutation [11, 21, 23]. The

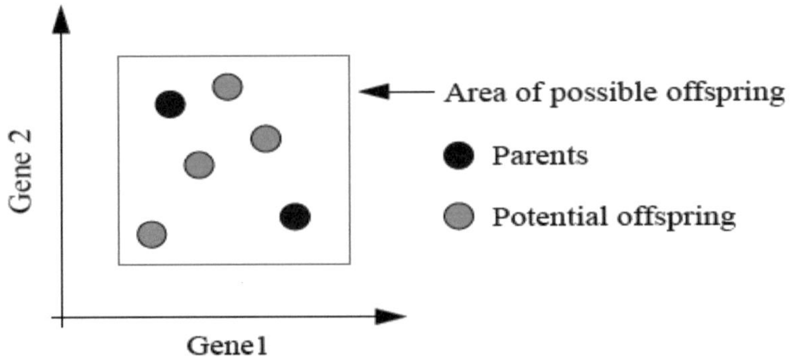

Figure 1.
Volume recombination.

diversity in the population is preserved by adding small, normally distributed zero-mean random numbers to each child before inserting it into the population. The random numbers have a certain standard deviation R [18]. The value of R should be selected carefully because it is critical in determining the convergence of the optimization. If the value of R is too small, the solution might result in premature convergence, while a high value of R might be detrimental to the optimal convergence of the algorithm [11, 23]. The adaptive mutation method proposed here allows us to determine the appropriate value of R. To achieve this, the population is divided into two halves, P1 and P2. P1 is assigned a mutation rate of double R (2R), while P2 is assigned a mutation rate of half R (R/2). The mutation rate R is adjusted depending on the performance of each half of the population (P1 or P2). If P1 gives better and fitter individuals, the mutation rate is increased by a certain percentage (10% in this case); similarly, if P2 produces better and fitter individuals, then the mutation rate gets reduced by a similar percentage. The pseudo code for BGA with adaptive learning can be found in [11, 23].

3. Overview of population-based incremental learning algorithm

Population-based incremental learning (PBIL) is a combination of competitive learning derived from artificial neural networks and genetic algorithms [18, 19]. There is no crossover operator in PBIL, instead, the probability vector is updated using a solution with the highest fitness values [18]. The values of the probability vector are initially set to 0.5 to ensure that the probability of generating 0 or 1 is equal. As the search progresses, these values are moved away from 0.5, toward either 0.0 or 1.0.

3.1 Learning rate

Learning in PBIL is based on using the current probability distribution to create N individuals. The probability vector is updated using the best individual so far, thereby increasing the probability of producing solutions similar to the current best solutions. Learning rate is required to update the probability vector. The selection of the learning rate value should be made with care as it determines how fast or slow the prototype vector is shifted toward the best individuals. A larger rate speeds up convergence, but it reduces the function space to be searched, while a smaller rate will slow down the convergence, even though it increases the exploration of a bigger search space, thereby increasing the likelihood of better optimal solutions. The (positive) update rule of the probability vector is given as:

$$PV_i = (1 - LR)PV_i + (LR)B_i \tag{3}$$

where PV is the probability vector, LR ∈ [0 1] is the learning rate, B is the best solution, and i denotes each locus (i = 1, 2, ... l) where l is the binary encoding length.

3.2 Mutation

Like in GA, the mutation is used in PBIL to maintain diversity in the population. Mutation in PBIL can be performed in two ways: either on the sample solutions generated or on the PV. In this study, the mutation is performed on the PV; a forgetting factor is used to relax the probability vector toward a neutral value of 0.5 [11, 16, 17] as shown in the equation below.

$$PV_i = PV_i - FF(PV_i - 0.5) \tag{4}$$

where *FF* is the forgetting factor that was chosen to be 0.005.

The pseudo code for PBIL can be found in [17–19].

4. Overview of differential evolution

Differential evolution (DE) can be defined as a parallel direct search method that uses a population of points to search for a global minimum or maximum of a function over a wide search space [13]. It is a simple and efficient adaptive scheme for global optimization over continuous space. DE is designed to efficiently solve non-differentiable and nonlinear functions and yet retains its simplicity and good convergence to a global optimum [12]. Similar to most EAs, DE explores the search space by maintaining a population of candidate solutions and by using Darwinian evolution theory to direct its search toward prospective areas. The candidates with better fitness values survive and enter the next generation [12–14, 25]. The process continues until the termination criterion is satisfied. It should be mentioned that DE has proved to be one of the best among EAs. It was able to secure competitive rankings in CEC competitions [25]. One of the main advantages of DE over GA is the mutation scheme and the selection process. Unlike GAs where the best solutions are selected for the next generation, in DE, all solutions have an equal chance of being selected as parents independently of their fitness values.

4.1 Mutation

In the context of DE, "mutation" is defined as a process of taking a small random sample of vectors from the current population and combining them algebraically to form a new vector, which is referred to as a *mutant vector* [12, 13]. In the so-called classical version of DE, the mutant vector is formed as follows:

$$V_{i,g} = X_{r1,g} + F\left(X_{r2,g} - X_{r3,g}\right) \tag{5}$$

where i, $r1$, $r2$, and $r3$ are all distinct indices in the interval $[1, N_p]$. The mutation scale factor F is a positive real number between 0 and 2 that controls the rate at which the population evolves [13]. The vector X_{r1} is the *base vector*, while $X_{r2} - X_{r3}$ is the *difference vector*, $g = 0, 1, \ldots g_{max}$ are the generations and N_p is the population.

The above process is repeated N_p-times to constitute a mutant population. In the classical version, each base vector is used only once per generation, in order to preserve diversity in the population. The classical version described above is designated as "DE/rand/1" and is widely used, although it has the drawback of relatively slow convergence [12]. Some alternative mutation strategies to the classical version are given below [12–14, 25]:

DE/best/1: This strategy resembles DE/rand/1, except that all mutants use the best vector in the current generation as the base vector:

$$V_{i,g} = X_{best,g} + F\left(X_{r1,g} - X_{r2,g}\right) \tag{6}$$

where X_{r1} and X_{r2} are distinct random vectors and X_{best} is the best individual in the current population.

This strategy has faster convergence than DE/rand/1, but often fails to reach the global optimum [12].

DE/best/2: This strategy uses two mutation differences to create a mutant vector:

$$V_{i,g} = X_{best,g} + F\left(X_{r1,g} - X_{r2,g}\right) + F\left(X_{r3,g} - X_{r4,g}\right) \tag{7}$$

where X_{r1}, X_{r2}, X_{r3}, and X_{r4} are distinct random vectors and X_{best} is the best individual in the current population. This strategy attempts to balance between convergence speed and robustness. However, it may still converge to a local but non-global optimum due to the fact that the base vector X_{best} draws the population toward itself [13].

DE/local-to-best/2: This strategy resembles DE/best/2 in that two mutation differences are used, but the base vector is randomly sampled and the "best" vector is used in one of the scaled differences:

$$V_{i,g} = X_{r1,g} + F\left(X_{best,g} - X_{r2,g}\right) + F\left(X_{r3,g} - X_{r4,g}\right) \tag{8}$$

This approach has similar convergence properties to DE/best/2 [13].

DE/rand/2: This strategy samples 5 random vectors in the current generation to form two random differences that are scaled and added to the base vector:

$$V_{i,g} = X_{r1,g} + F\left(X_{r2,g} - X_{r3,g}\right) + F\left(X_{r4,g} - X_{r5,g}\right) \tag{9}$$

where, $r1 \neq r2 \neq r3 \neq r4 \neq r5$. This approach converges more slowly but is very robust [13].

DE/rand/2 has been used in this work due to our objective to appropriately tune the PSS with optimal time constants values for a robust performance.

4.2 Crossover

In DE, "crossover" refers to the process of creating a new vector (called the *trial vector*) by combining a mutant vector with a *target vector* [13]. The target vector for the mutant vector $V_{i,g}$ is $X_{i,g}$. The trial vector $U_i = [u_{1,i}, u_{2,i} \dots, u_{D,i}]$, is then obtained as follows:

$$u_{j,i,g} = \begin{cases} v_{j,i,g} \ \textit{if} \left(rand_j(0,1) \le CR \ or \ j = j_{rand}\right), \ j = 1, 2, \dots, D \\ x_{j,i,g} \ \textit{otherwise} \end{cases} \tag{10}$$

where $CR \in [0, 1]$ is the *crossover probability,* and CR is the fraction of the parameter values that are copied from the mutant vector, and 1-CR is the fraction of parameter values copied from the trial vector. To determine whether the parameter to be copied is from the mutant or trial vector, a uniformly-distributed random number, $rand_j$ between [0, 1] is generated and compared to the predefined value of CR. In addition, a random index $j_{rand} \in [1, N_p]$ is chosen and the corresponding mutant parameter is copied to ensure that the trial vector is not a duplicate of the target vector.

4.3 Selection

This process consists of choosing the individuals that will enter the next generation. DE employs a "one-to-one survivor selection," which consists of comparing each trial vector to its corresponding target vector. Mathematically, the vector $X_{i,g+1}$ in the g + 1'th generation is obtained from the trial vector $U_{i,g}$ and target vector $X_{i,g}$ as follows in the case of a minimization problem:

$$X_{i,g+1} = \begin{cases} U_{i,g} \text{if} \, f(U_{i,g}) \leq f(X_{i,g}) \\ X_{i,g} \quad otherwise \end{cases} \tag{11}$$

This process ensures that the best vector at each index is retained. Furthermore, this also guarantees that the very best-so-far solution is kept. Once the selection is performed for all target vectors in the current generation g, the processes of mutation, crossover, and selection are repeated with the N_p vectors in the $g + 1$st generation. This process is iterated until a termination criterion is satisfied.

5. System model

The power system considered in this paper is the two-area four-machine power system as shown in **Figure 2** [1]. Each machine is represented by the detailed six-order differential equations. The machines are equipped with simple exciter systems of first-order differential equations as given in the **Appendix** [11]. The system consists of two similar areas connected by a tie-line. Each area consists of two coupled conventional generator units, each generator is rated 900 MVA and 20 kV. The generator parameters can be found in [1, 11]. The dynamics of the system are described by a set of nonlinear differential equations. However, for the purpose of controller design, these equations are linearized around the nominal operating conditions. The linearized equation of the system is given by:

$$\begin{aligned} x &= A_o x + B_o u \\ y &= C_o x + D_o u \end{aligned} \tag{12}$$

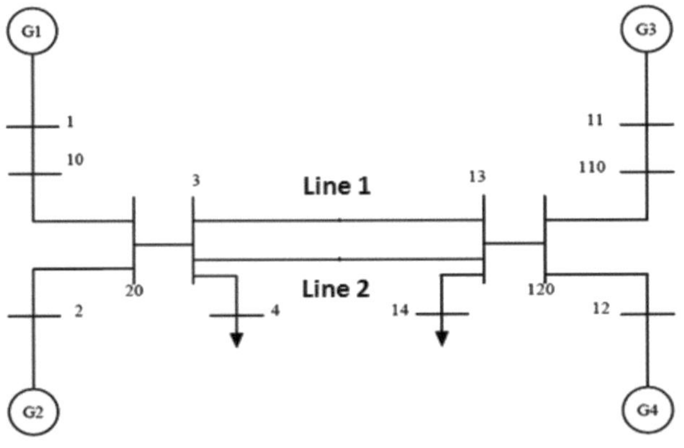

Figure 2.
Two-area system model.

where, x is the state variable, y is the system output, and u is the control input. A_O, B_o, C_o, and D_o are constant matrices of appropriate dimensions.

Several operating conditions have been considered during the design stage of the controller. However, only three operating conditions are listed in **Table 1** for simplicity. Case 1 is the nominal operating condition. At the nominal operating condition, approximately 146 MW is transferred from area 1 to area 2 via the two tie lines, with each line carrying half of the total power. Under these conditions, the load on bus 4 was 1137 MW, while the load on bus 14 was 1367 MW. Case 2 is the moderate load condition, where about 409 MW of real power is transferred from area 1 to area 2. For this case, the load on bus 4 was 967 MW, while the load on bus 14 was 1767 MW. case 3 is the heavy load condition (worst case scenario) where approximately 512 MW of power is transferred from area 1 to area 2. For this case, the load on bus 4 was 876 MW, while the load on bus 14 was 1876 MW. It should be mentioned that the system exhibits inter-area oscillatory modes due to the flow of power between the two areas that causes the two areas to oscillate against each other. In addition, two local area modes were also observed, one in each area. However, in this chapter, we will concentrate only on the inter-area modes since they are the most critical and difficult to control. **Table 2** shows the open-loop eigenvalues of the inter-area modes. It can be seen that without PSSs, the inter-area modes were stable but poorly damped for case 1, with a damping ratio of 0.011. However, the system became unstable for case 2 and the instability became more pronounced for case 3 with damping ratios of -0.0057 and $-$ 0.0130, respectively. This suggests that with the increase in active power transfer between the two areas, the oscillations have now increased making the system unstable. The frequency of oscillations of the inter-area modes ranges from 0.588 Hz to 0.634 Hz.

Therefore, a supplementary controller known as a power system stabilizer (PSS) will be required to damp the system's oscillations. The block diagram of the PSS is shown in **Figure A.1** in the **Appendix**.

6. Objective function

The objective is to optimize the parameters of the PSSs simultaneously such that the controllers can stabilize the system over a wide range of operating conditions. The parameters that were to be optimized are K (gain of the PSS) as well as the lead-lag time constants T_1, T_2, T_3, and T_4. The objective function used was to maximize the lowest damped ratio over a wide range of operating conditions. This objective function was used for GA, BGA, PBIL, and DE. The objective function is given as:

Case	Active power transfer from area 1 to area 2 [MW]	Number of tie-line between areas 1 and 2	Load's active power at bus 4 [MW]	Load's active power at bus 14 [MW]
1	146	2	1137	1367
2	409	2	967	1767
3	512	2	876	1876

Table 1.
Selected operating conditions.

Case	Inter-area mode	Damping ratio (%)	Frequency of oscillations (Hz)
1	$-0.044 \pm j3.98$	1.10	0.634
2	$0.022 \pm j3.78$	−0.57	0.602
3	$0.048 \pm j3.69$	−1.30	0.588

Table 2.
Open-loop eigenvalues of the inter-area modes for selected operating conditions.

$$val = \max\left(\min\left(\varsigma_{ij}\right)\right) \tag{13}$$

where
$i = 1,2, \ldots n, j = 1, 2, \ldots .m$

$$\varsigma_{ij} = \frac{-\sigma_{ij}}{\sqrt{\sigma_{ij}^2 + \omega_{ij}^2}} \tag{14}$$

where $\zeta_{ij} = \frac{-\sigma_i}{\sqrt{\sigma_i^2+\omega_i^2}}\, j$ is the damping ratio of the *i*th eigenvalue of the jth operating conditions. The number of the eigenvalues is n, and m is the number of operating conditions. σ_{ij} and ω_{ij} are the real part and the imaginary part (frequency) of the eigenvalue, respectively.

7. Design of the PSSs

In total 10 PSSs parameters were optimized (i.e., 5 parameters for each area) for generators 1–4. The parameters that were optimized are K, T_1, T_2, T_3, and T_4. The washout time constant (T_w) was set at 10 seconds and was not optimized since T_w is not critical to the design. The following parameter domain constraints were considered when designing the PSSs.

$$0 < K \le 20$$

$$0.001\, T_i \le 5$$

where K and T_i (i = 1, 2, 3, 4) denote the controller gain and the lead–lag time constants, respectively.

For comparison purposes, a CPSS was also designed using the phase compensation technique. Details can be found in [1, 2].

7.1 Parameters of GAs, BGA, PBIL, and DE

The parameters used in the optimization for GAs, BGA, PBIL, and DE are shown in **Table 3**.

An observation of the parameters given below in **Table 3** shows that PBIL uses few parameters. There is no crossover or selection in PBIL compared to BGA, GA, and DE. In addition, 500 generations were used in the PBIL optimization to allow for adequate learning to take place within the optimization. This is because PBIL that works by learning from the previous best and trying to find the very best individual takes time to explore the

Parameters	GA	BGA	PBIL	DE
Population	100	100	100	50
Generation	120	120	500	180
Selection	Normal geometric	Truncation selection	—	Greedy
Crossover/ Recombination	Arithmetic	Line and volume	—	Binomial (CR: 0.95)
Mutation	Nonuniform	Adaptive random (initial R_{nom}: 0.01)	Forgetting Factor (FF:0.005)	DE/rand/2 (F: 0.95)
Learning rate (LR)	—	—	0.1	—

Table 3.
Parameters used in GA, BGA, PBIL, and DE.

search space. Another difference is the way in which the initial population is generated. In GA, BGA, and DE, the initial population is selected randomly, while in PBIL the role of the population is redefined using probability vectors (PV). It should be mentioned that a population size of 50 was also tested in PBIL and it was found that it yielded similar results to the population size of 100. However, in this work a population of 100 was used.

7.2 Conventional PSS

The parameters of the conventional PSS (CPSS) were tuned at the nominal operating condition using the phase compensation method and trial and error approach. Details of this approach can be found in [1–3].

8. Simulation results

8.1 Fitness values

Figures 3–6 show the fitness value (minimum damping ratio) of the system when GA, BGA, PBIL, and DE are used in the optimization. The final value obtained from the GA optimization is 0.1867 as compared to 0.205, 0.2095, and 0.227 for BGA, PBIL, and DE, respectively. As discussed previously, GA and BGA were run for 120 generations, DE for 180 generations, while the PBIL was run for 500 generations. Since a smaller population was used for DE, it was decided to increase its generations. The reason for using 500 generations in PBIL is that it starts to settle only around 300 generations and therefore there is a need for a longer simulation period.

8.2 Eigenvalue analysis

Table 4 shows the inter-area modes for the system with the PSSs. It can be seen that with the PSSs, the inter-area modes are very well damped as compared to the open-loop system in **Table 2**. CPSS performs adequately for the nominal operating condition. The damping ratios provided by the CPSS under the three cases 1, 2, and 3, are 0.1666, 0.1442, and 0.1339, respectively. BGA-PSS provides a damping ratio of 0.2321, 0.2393, and 0.2412 for cases 1, 2, and 3, respectively. On the other hand, the

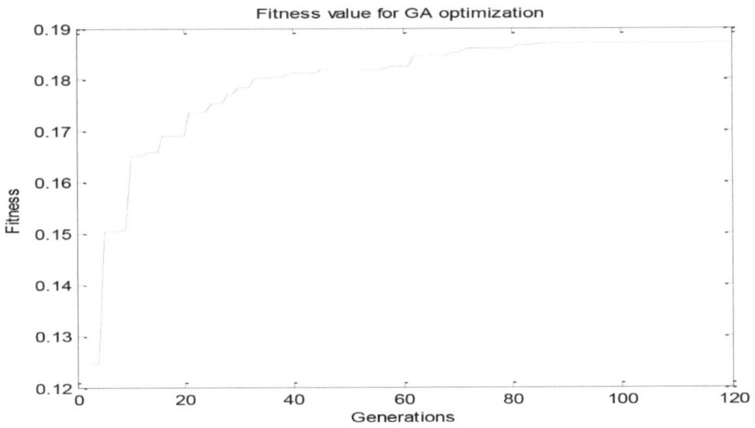

Figure 3.
Fitness value curve from the GA optimization.

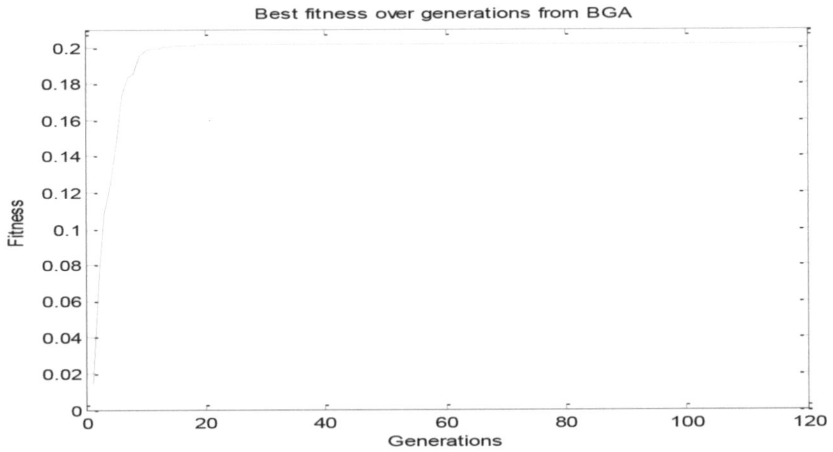

Figure 4.
Fitness value curve from the BGA optimization.

PBIL-PSS and DE-PSS provide a damping ratio of 0.2341 and 0.2377, respectively, for case 1; 0.2387 and 0.2321, respectively, for case 2; 0.2385 and 0.23, respectively, for case 3.

It is observed that PBIL-PSS, DE-PSS, and BGA-PSS provide similar damping ratios to the system for operating condition considered. In case 1, DE provides the best damping ratio, whereas BGA provides the best damping ratios for cases 2 and 3. Among the evolutionary algorithm-based PSSs, GA-PSS provides the lowest damping ratios of 0.2029, 0.2013, and 0.1993 for cases 1, 2, and 3, respectively.

Figure 7 shows the spread of the eigenvalues for the system equipped with the different PSSs. CPSS is the lowest compared to the damping provided by all the other EA-based PSSs. It is observed that among the EA-based PSSs, GA-PSS provides the least damping. The damping provided by the PBIL-PSS, BGA-PSS, and DE-PSS is very similar and higher than that provided by GA-PSS.

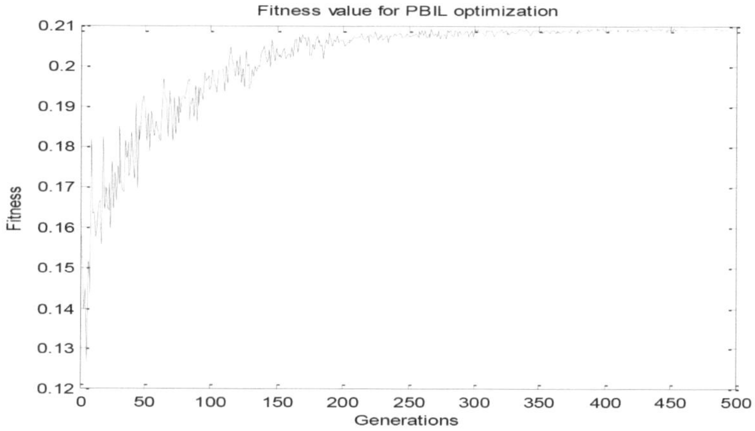

Figure 5.
Fitness value curve from the PBIL optimization.

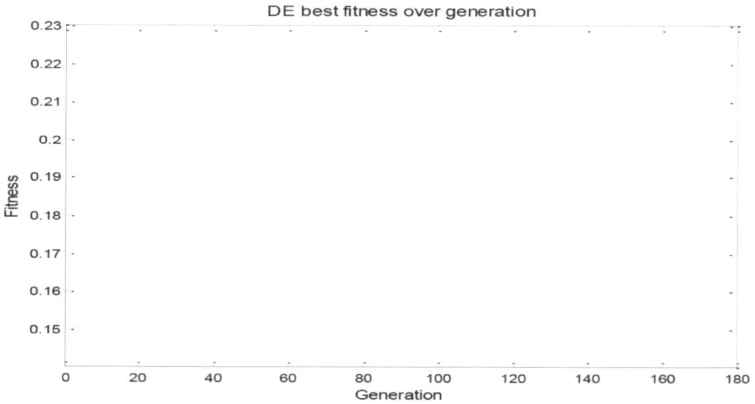

Figure 6.
Fitness value curve from DE optimization.

Case	CPSS	GA-PSS	BGA-PSS	PBIL-PSS	DE-PSS
1	$-0.62 \pm j3.67$ (0.1666)	$-0.80 \pm j3.86$ (0.2029)	$-0.89 \pm j3.73$ (0.2321)	$-0.91 \pm j3.78$ (0.2341)	$-0.94 \pm j3.84$ (0.2377)
2	$-0.50 \pm j3.43$ (0.1442)	$-0.75 \pm j3.65$ (0.2013)	$-0.86 \pm j3.49$ (0.2393)	$-0.87 \pm j3.54$ (0.2387)	$-0.89 \pm j3.73$ (0.2321)
3	$-0.45 \pm j3.33$ (0.1339)	$-0.72 \pm j3.54$ (0.1993)	$-0.84 \pm j3.38$ (0.2412)	$-0.84 \pm j3.42$ (0.2385)	$-0.87 \pm j3.68$ (0.2300)

Table 4.
Inter-area modes and the respective damping ratios in brackets.

8.3 Small disturbance

To investigate the performance of the PSSs under small disturbance, a small disturbance of 5% step response is applied to the reference voltage of generator 2 in area 1. The responses of the active power output of generators 2 and 3 are are shown in

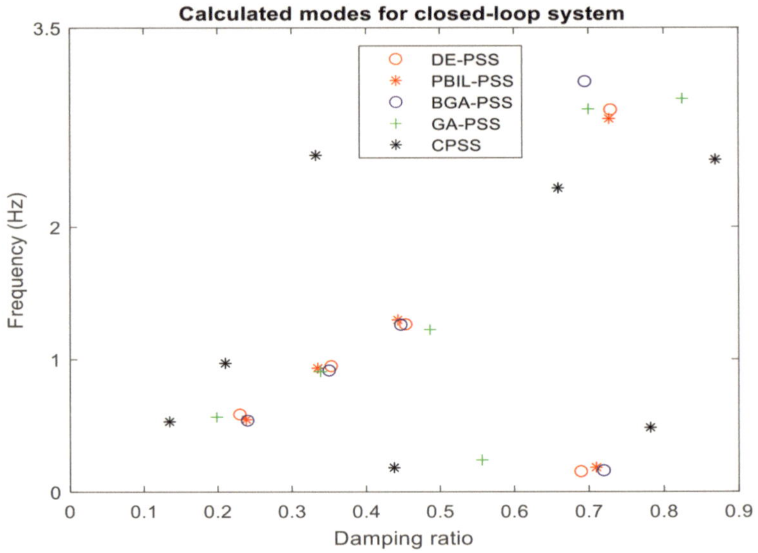

Figure 7.
Spread of the eigenvalues for the different PSSs-nominal condition.

Figures 8–13 for cases 1, 2, and 3, respectively. It can be seen that the system is well-damped across all three operating conditions when it is equipped with DE-PSS, BGA-PSS, GA-PSS, and PBIL-PSS. The CPSS is seen to give the worst performance.

Figures 8 and **9** show the active power output responses of generators 2 and 3, respectively, for case 1. The system equipped with GA-PSS, BGA-PSS, DE-PSS, and

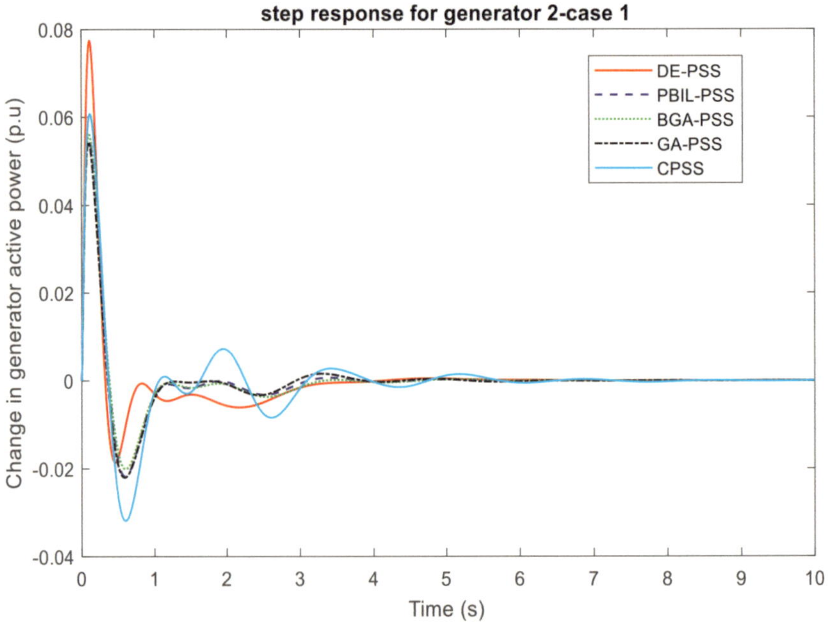

Figure 8.
Response of G2 under the 5% step change in Vref of G2 – Case 1.

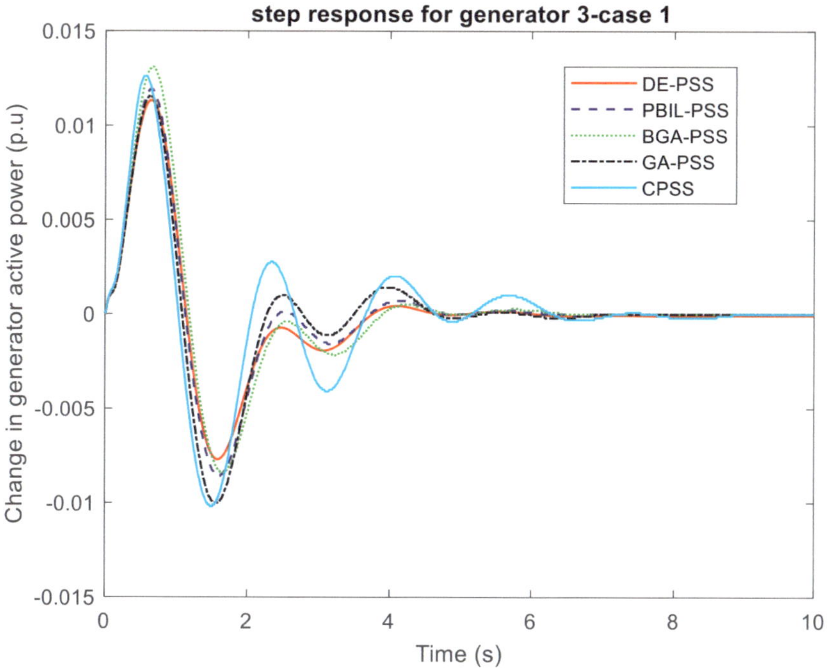

Figure 9.
Response of G3 under the 5% step change in Vref of G2 – Case 1.

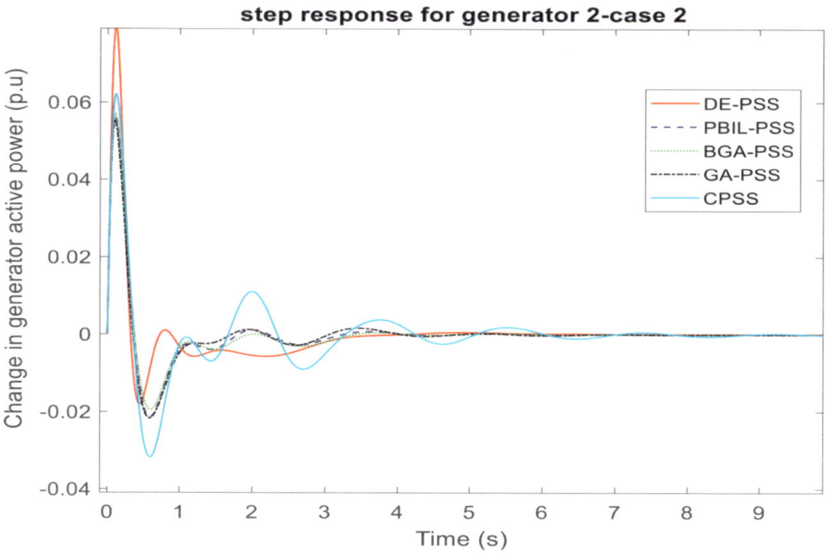

Figure 10.
Response of G2 under the 5% step change in Vref of G2 – Case 2.

PBIL-PSS has a similar settling time of approximately 4 sec., whereas the system equipped with CPSS has a longer settling time of around 6 sec. DE-PSS is seen to give the best performance in terms of undershoot and the amplitude of subsequent swings, albeit with a relatively large 1st swing overshoot as seen in **Figure 8**. It is observed that DE-PSS gives a large 1st swing overshoot in **Figure 8**. The relatively large 1st swing overshoot can be attributed to the high gain of the controller. Note that DE-PSS's gain has almost reached the allowable maximum value [20]. The performance of BGA-PSS is comparable to that of PBIL-PSS. Compared with other EA-based PSS, GA-PSS gives the worst performance. However, it performed better than the CPSS. In **Figure 9**, BGA-PSS is seen to give a slightly high 1st swing overshoot but the subsequent swings are well-damped. Overall, CPSS is seen to give the worst performance.

Figures 10 and **11** show the active power responses of generators 2 and 3, respectively, for case 2. It can be seen that the CPSS has a longer settling time of around 7 sec. Compared to a settling time of around 4 sec. for the EA-based PSSs. This suggests that the oscillations have increased in case 2 compared to case 1. The EA-based PSSs are able to damp the oscillations adequately when compared to the CPSS. In terms of undershoot and subsequent swings, DE-PSS is seen to give the best responses albeit with a relatively large 1st swing overshoot as seen in **Figure 10**. The performances of BGA-PSS and PBIL-PSS are similar. Overall, CPSS gives the worst performance followed by GA-PSS.

Figures 12 and **13** show the active power responses of generators 2 and 3, respectively, for case 3. It can be seen that the system response is similar to case 2 except that the oscillations have now increased as can be seen in the system's responses. The system equipped with the CPSS settled around 10 sec. (see **Figure 13**). It can be seen that the performance of the CPSS has now deteriorated significantly. On the other hand, the performances of GA-PSS, BGA-PSS, PBIL-PSS, and DE-PSS have deteriorated only slightly. This means that the EA-based PSSs are more robust. In terms of settling time, the EA-based PSSs have similar settling times of approximately 6.5 sec., which is comparable to case 2. Although DE-PSS has a larger 1st swing overshoot as seen in **Figure 12**, it gave the best responses in terms of undershoot and subsequent swing amplitudes, followed by BGA-PSS and PBIL-PSS. The performance of GA-PSS although better than that of CPSS is not as good as the other EA-based PSS.

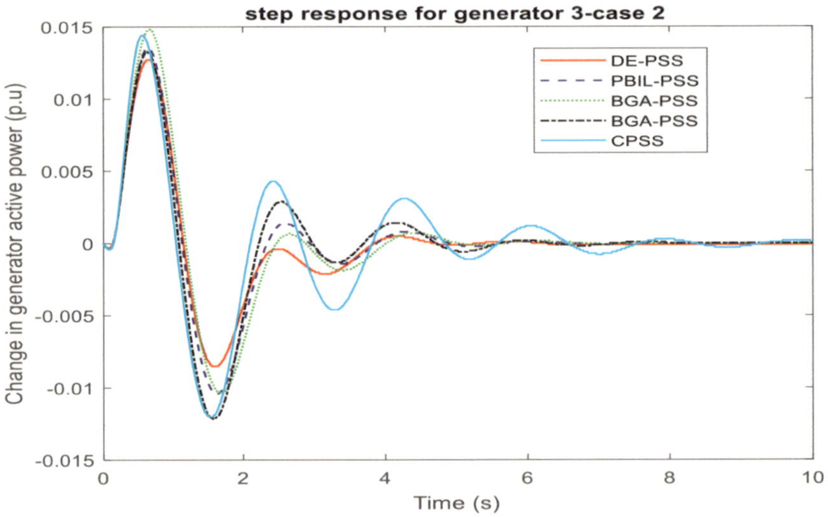

Figure 11.
Response of G3 under the 5% step change in Vref of G2 – Case 2.

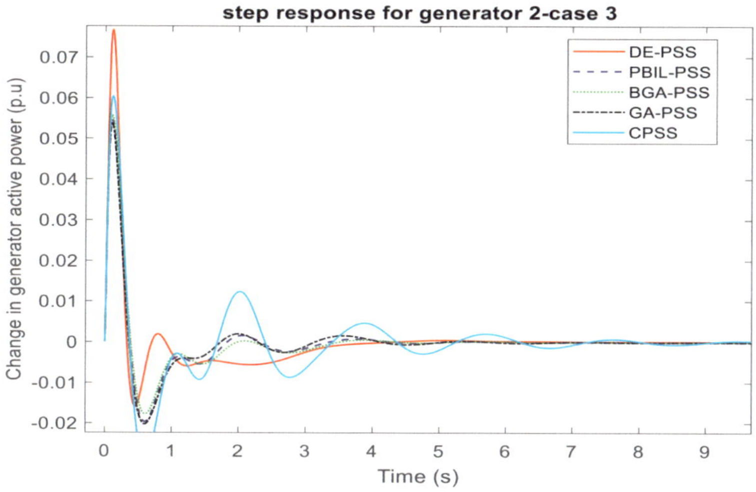

Figure 12.
Response of G2 under the 5% step change in Vref of G2 – Case 3.

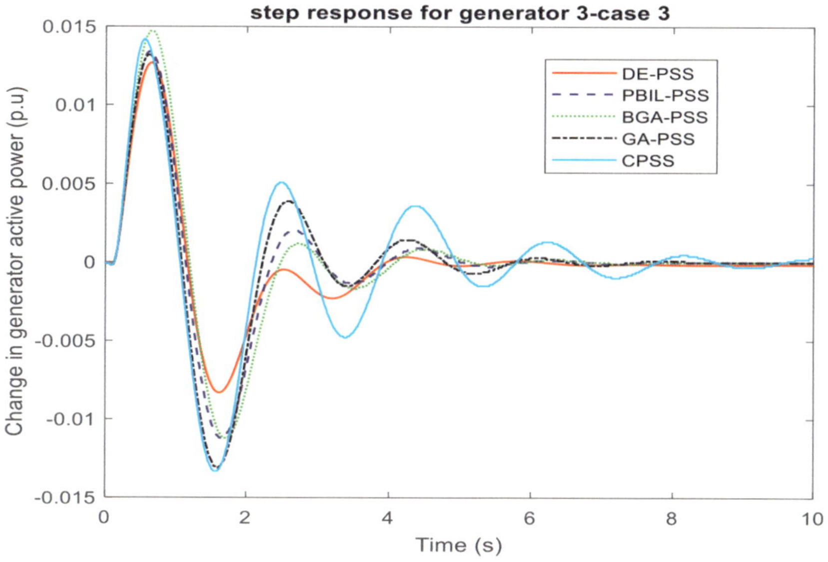

Figure 13.
Response of G3 under the 5% step change in Vref of G2 – Case 3.

8.4 Large disturbance

A large disturbance was considered by applying a three-phase fault to the system at bus 3. The fault was cleared by removing one of the transmission lines between bus 3 and bus 13. The fault was applied for 0.1 seconds. After the fault was cleared, the faulted line was removed and the system settled to a different operating condition with only one tie line transmitting power from area 1 to area 2. This means the system is weaker after the fault was cleared compared to its state before the fault. **Figures 14** and **15** show

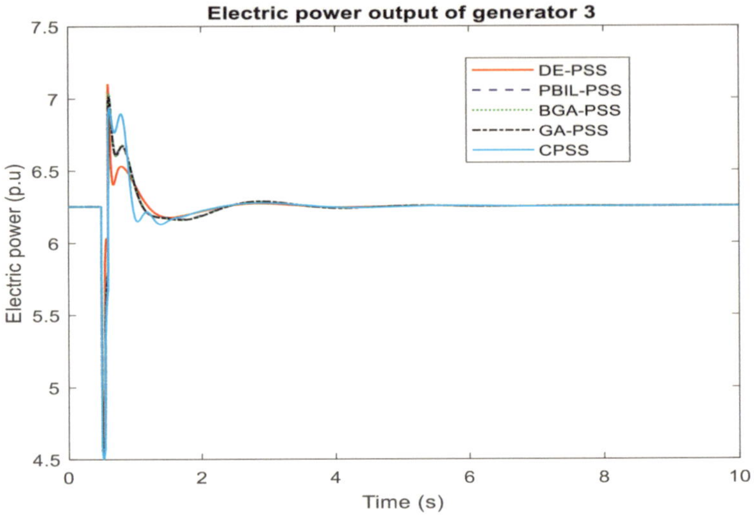

Figure 14.
Electric power output of generator 3 following a three-phase fault on bus 3 for case 1.

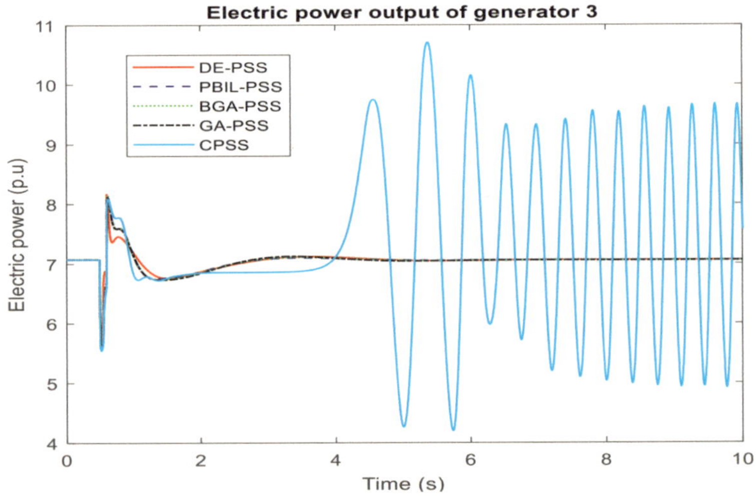

Figure 15.
Electric power output of generator 3 following a three-phase fault on bus 3 for case 2.

the electric power output for generator 3 for case 1 and case 2, respectively. The responses for case 3 are not shown because the system was unable to survive this large disturbance after the fault was removed. It can be seen from **Figure 14** (case 1) that the output power of generator 3 has a high overshoot in the first swing after the fault was cleared but settled down quickly after a few seconds, with all the PSSs providing adequate damping to stabilize the system. However, when the power that was transferred from area 1 to area 2 increased,the CPSS was unable to maintain the stability of the system as seen in **Figure 15** (case 2). On the other hand, all the EA-based PSSs were able to stabilize the system, which suggests that they are more robust than the CPSS.

9. Conclusions

An optimal PSS design for small signal stability improvement of a multi-machine power system using four evolutionary algorithms (GA, BGA, PBIL, and DE) has been presented. Frequency-domain and time-domain simulations have been presented to show the effectiveness of the EA-based PSSs in damping low-frequency oscillations. It is shown that in the frequency domain, the performances of BGA-PSS, PBIL-PSS, and DE-PSS are comparable and better than that of the GA-PSS for all cases investigated. However, time-domain simulations show that DE-PSS performs better than BGA-PSS and PBIL-PSS in terms of undershoot and subsequent swings albeit with a relatively large 1st swing overshoot. This overshoot could be attributed to the high gain of the controller. One way to deal with this overshoot is to reduce the gain of the controller; however, this could also affect the damping. GA-PSS is shown to give the worst performance among the EA-based PSSs, but it performed better than the CPSS. In designing the PBIL-PSS, more generations were required compared to GA-PSS, BGA-PSS, and DE-PSS. Since PBIL works by learning from the previous best individual, it takes time for the algorithm to explore the search space. Compared to the EA-based PSS, the CPSS that was designed using the conventional method has been shown to perform poorly and is not robust. Further research will be done in the direction of improving the EAs algorithms by self-adapting the genetic parameters and using multi-objective functions in the optimization.

Acknowledgements

This research was funded in part by the National Research Foundation (NRF) of South Africa, Grant UID 118550.

Appendix

Generator and automatic voltage regulator (AVR) equations

$$\frac{d}{dt}E_{fd} = \frac{K_A}{T_A}\left(V_{ref} - V_t\right) - \frac{E_{fd}}{T_A}$$

where K_A and T_A are the gain and time constant of the AVR. V_t is the terminal voltage of the generator. In this work, K_A = 200 and T_A = 0.05 sec.

PSS block diagram

where K is the gain of the PSS, T_1 to T_4 are lead/lag time constants, and Tw is the washout time constant. T_1 and T_2 form the first lead/lag block, while T_3 and T_4 form the second lead/lag block of the PSS.

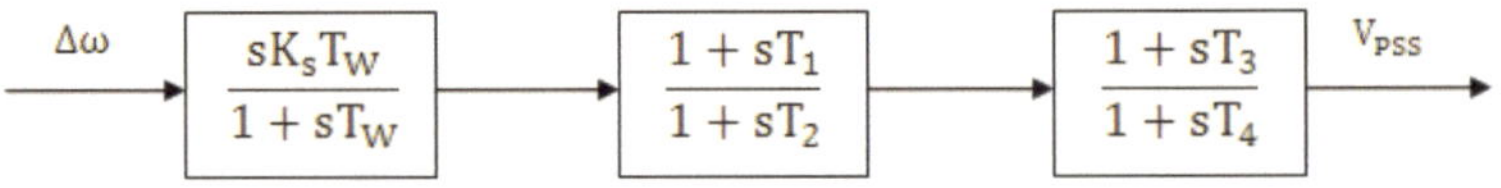

Figure A.1.
PSS block diagram.

Author details

Komla Agbenyo Folly*, Severus Panduleni Sheetekela and Tshina Fa Mulumba
University of Cape Town, South Africa

*Address all correspondence to: komla.folly@uct.ac.za

References

[1] Kundur P. Power System Stability and Control. USA: Prentice-Hall; 1994

[2] Klein M, Rogers GJ, Kundur P. A fundamental study of inter-area oscillations in power systems. IEEE Transactions on Power Systems. 1991; **6**(3):914-921

[3] Chen L. A Novel Method for Power System Stabilizer Design. Cape Town, South Africa: University of Cape Town; 2003

[4] Du W, Dong W, Wang Y, Wang H. A method to design power system stabilizers in a multi-machine power system based on single-machine infinite-bus model. IEEE Transaction on Power Systems. 2021;**36**(4):3475-3486. DOI: 10.1109/TPWRS.2020.3041037

[5] Chow JH, Sanchez-Gasca JJ. Power system stabilizers. In: Power System Modeling, Computation and Control. 2020. pp. 265-294. DOI: 10.1002/9781119546924.ch10

[6] Folly KA, Yorino N, Sasaki H. Improving the robustness of H∞-PSSs using the polynomial approach. IEEE Transactions on Power Systems. 1998; **13**(4):1359-1364

[7] Holland JH. Adaptation in Nature and Artificial Systems. Ann Arbor: University of Michigan Press; 1975

[8] Goldberg DE. Genetic Algorithms in Search, Optimization & Machine Learning. USA: Addison-Wesley; 1989

[9] Mitchell M. An Introduction to Genetic Algorithms. Cambridge MA, United States: The MIT Press; 1996

[10] Alkhatib H, Duveau J. Robust design of power system stabilizers using adaptive genetic algorithms. In: Proceeding of the Word Academy of Science, Engineering, and Technology. 2010. pp. 267-272

[11] Sheetekela S. Design of Power System Stabilizer using Evolutionary Algorithms. Cape Town, South Africa: University of Cape Town; 2010

[12] Mulumba TF, Folly KA. Application of evolutionary algorithms to power system stabilizer design. In: Subair S, Thron C, editors. Implementation and Application of Machine Learning. Studies in Computational Intelligent (SC 782). 2020. pp. 29-62

[13] Price K, Storn R, Lampinen J. Differential Evolution—A Practical Approach to Global Optimization. Berlin, Germany: Springer; 2005

[14] Ahmad MF, Isa NAM, Lim WH, Ang KM. Differential evolution: A recent review based on state-of-the-art works. Alexandria England Journal. 2022;**61**: 3831-3872

[15] Verdejo H, Pino V, Kliemann W, Becker C, Delpiano J. Implementation of particle swarm optimization (PSO) algorithm for tuning power system stabilizers in multi-machine electric power systems. Energies. 2020;**13**(8): 2093. DOI: 10.3390/en13082093

[16] Folly KA. Performance of power system stabilizers based on population-based incremental learning (PBIL) algorithm. International Journal of Electrical Power and Energy System. 2011;**33**(7):1279-1287

[17] Folly KA. Parallel PBIL applied to power system controller design. Journal of Artificial Intelligence and Soft

Computing Research. 2013;**3**(3): 215-223. DOI: 10.2478/jaiscr-2014-0015

[18] Baluja S. Population-Based Incremental Learning: A method for integrating Genetic Search Based Function Optimization and Competitive Learning. Technical Report CMU-CS-49-163, 1994

[19] Baluja S, Caruana R. Removing the genetics from the standard genetic algorithm. In: Proceedings of the 12th International Conference on Machine Learning, Lake Tahoe, CA; 1995

[20] Sheetekela S, Folly KA. Multimachine power system stabilizer design based on evolutionary algorithm. In: Proceedings of the 44th International Universities' Power Engineering Conference. 2009

[21] Sheetekela S, Folly KA.: Breeder genetic algorithm for power system stabilizer design. In: Proceedings of 2010 IEEE Congress on Evolutionary Computation (CEC), Barcelona, Spain; 2010

[22] Mühlenbein H, Schlierkamp-Voosen D. Predictive models for the Breeder Genetic Algorithm, I. continuous parameter optimization. Evolutionary Computation. 1993;**1**(1):25-49

[23] Greene J. The Basic Idea behind the Breeder Genetic Algorithm. Cape Town, South Africa: University of Cape Town; 2005

[24] Folly KA, Sheetekela SP. Optimal design of power system controller using breeder genetic algorithm. In: Gao S, editor. Bio-Inspired Computational Algorithms and Their Applications. InTech-open science; 2012. pp. 303-316. DOI: 10.5772/38447

[25] Das S, Suganthan PN. Differential evolution: A survey of the state-of-the-art. IEEE Transaction on Evolutionary. Computation. 2011;**15**(1):4-31

Chapter 4

ADDC: Automatic Design of Digital Circuit

Conor Ryan, Michael Tetteh, Jack McEllin, Douglas Mota-Dias, Richard Conway and Enrique Naredo

Abstract

Digital circuits are one of the most important enabling technologies in the world today. Powerful tools, such as Hardware Description Languages (HDLs) have evolved over the past number of decades to allow designers to operate at high levels of abstraction and expressiveness, rather than at the gate level, which circuits are actually constructed from. Similarly, highly accurate digital circuit simulators permit designers to test their circuits before committing them to silicon. This is still a highly complex and generally manual task, however, with complex circuits taking months or even years to go from planning to silicon. We show how Grammatical Evolution (GE) can harness the standard tools of silicon design and be used to create a fully automatic circuit design system. Specifically, we use a HDL known as SystemVerilog and Icarus, a free, but powerful simulator, to generate circuits from high level descriptions. We apply our system to several well known digital circuit literature benchmarks and demonstrate that GE can successfully evolve functional circuits, including several which have been subsequently rendered in Field Programmable Gate Arrays (FPGAs).

Keywords: digital design, VLSI design, microelectronics design, evolvable hardware, HDL, verilog, grammatical evolution

1. Introduction

This chapter describes the application of Evolutionary Computation to the task of digital circuit design. Although many Electronic Design Automation (EDA) tools exist to aid designers, digital circuit design remains a time consuming and expensive task that requires skilled engineers.

The cost of errors in silicon is enormous and this has led to extremely powerful and accurate simulators that designers use to verify their designs before committing them to silicon. These simulators provide a huge opportunity for Evolutionary Computation as they can be used to test individuals.

This chapter gives an overview about how digital integrated circuits are designed and how the tools used to develop them have evolved over the past few decades. These tools, when linked together with GE produce a system we call the Automatic Design of Digital Circuits (ADDC), which can evolve circuits using massive levels of abstraction rather than simple logic gates.

We demonstrate the system on three real-world problems, including one with 2^{20} test cases, showing that ADDC successfully evolves solutions in all cases.

2. Background

Digital circuit design began in the 1960's with the arrival of semiconductor transistor based circuits and the Integrated Circuit (IC). Up until the 2010's, *Moore's Law* has successfully predicted the shrinking in size of manufacturing technologies allowing for lower cost, faster speeds and lower power consumption. However in the last decade, this shrinking has slowed due to the difficulties involved in the fabrication process. Integrated circuits are extremely common in many products today and their use is not always obvious.

Integrated circuits come in three different varieties; Digital, Analogue or Mixed-Signal. Digital integrated circuits process digital information, often represented using bits, bytes or words. Many of these circuits employ the use of one or more processors (often referred to as a core) with support logic, memories and I/O interfaces. The microprocessor is a famous example of a digital circuit. Analogue integrated circuits are used for handling continuous-time signals and to perform operations such as amplification, analogue filtering and power management. Mixed-Signal integrated circuits contain both analogue and digital circuitry in the same package and use ADC (Analogue to Digital Converters) and DACs (Digital to Analogue Converters) to share information between both domains.

In modern circuit design, signal processing tends to be performed in the digital domain instead of the analogue domain. This is due to the reliability of digital circuitry and the existence of advanced digital algorithms with performance that cannot be obtained with analog circuitry alone [1]. This move towards using digital designs for signal processing has required the use of circuit representations like Hardware Description Languages (HDLs) to be used to describe these extremely complex circuits. New devices such as Complex Programmable Logic Devices (CPLDs) and Field Programmable Gate Arrays (FPGAs) are increasingly being used due to their ability to replicate the behaviour of these circuits without requiring the fabrication of new chips. The following sections will go more in-depth into HDLs, the differences between CPLD and FPGA devices and an overview of the Digital Design Flow.

2.1 Hardware description languages

The first modern HDL, Very High Speed Integrated Circuit Hardware Description Language (VHSIC-HDL), more commonly known as VHDL, was created in 1983. VHDL was developed for the US Department of Defense as part of the VHSIC project. The project was launched in 1980 [2], while the first version of VHDL was launched in 1983 by Intermetrics Inc., Texas Instruments and IBM [3, 4]. VHDL is a verbose and strongly-typed language. It grew steadily in popularity, resulting in both logic simulators and logic synthesis tools being developed for it. IEEE Standard VHDL was standardised in 1986 [5] with the adoption of VHDL version 7.2 and was finalised in 1987 in the IEEE Standard 1076-1987 [5]. VHDL would become the first HDL language that would gain widespread adoption, and is still in use today.

Another modern HDL developed around this time was Verilog, created by Phil Moorby in 1983 [6] while working for Gateway Design Automation, who were acquired by Cadence Design Systems in 1989 [7]. In comparison to VHDL, Verilog is less verbose and is a weakly-typed language. Originally it was designed only for logic simulation, but later had logic synthesis features added after the language gained widespread popularity. Verilog-XL, a Verilog simulator owned by Cadence, became the *de facto* Verilog simulator throughout the 1990's. Due to the increasing popularity of VHDL, Cadence created the the Open Verilog International (OVI, now known as Accellera) organisation and transferred the rights of Verilog into the public domain [8]. Verilog was eventually standardised in IEEE Standard 1364-1995 [6]. Verilog would be superseded by SystemVerilog in IEEE Standard 1800-2005 [9], adding features for design verification. SystemVerilog is more popular than VHDL today due to the language being less verbose and having a similar structure to the C programming language. **Table 1** provides a comparision between VHDL and SystemVerilog hardware description languages.

With the introduction of Hardware Description Languages for digital circuit design, two discrete time based simulation methods came into prominence. Both cycle-driven and event-driven simulation methods were orders of magnitude faster than the traditional continuous time based simulation method "SPICE". One limitation of the cycle-driven simulation method is that the output is only updated on each clock edge This means it can only be used for synchronous digital designs, but is much faster than event-driven simulation. It also cannot detect glitches and doesn't take the timing of the design into consideration.

Event-driven simulation updates the output on any input event meaning it can be used for both synchronous and asynchronous designs. Although still quicker than SPICE methods, it is much slower than cycle-driven simulation. Modern circuit designs utilise techniques such as clock and power gating, allowing parts of a design to be "turned off". This can help reduce the simulation time of an event-driven simulation, bringing it closer to cycle-driven simulation while providing a more accurate simulation. **Table 2** provides a comparison between cycle-driven and event-driven simulation methods. Practically all commercial and open-source simulation tools today utilise one of these methods.

2.2 CPLD vs FPGA devices

As digital designs grew in complexity, early Programmable Logic Devices (PLD) such as Programmable Array Logic (PAL) became obsolete as they could only replicate the behaviour of a few hundred logic gates. To address this shortcoming, PALs were soon replaced by Complex Programmable Logic Devices (CPLD). Modern CPLDs are able to replicate the behaviour of hundreds of thousands of logic gates.

VHDL	SystemVerilog
Standardised in 1987	Standardised in 1995 (Verilog) and 2005 (SystemVerilog)
More Verbose	Less Verbose
ADA-like	C-like
Case Insensitive	Case Sensitive
Support for Digital, Analog and Mixed-Signal Designs	Support for Digital Designs only

Table 1.
A comparison between VHDL and verilog hardware description languages.

Cycle-driven simulation	Event-driven simulation
Evaluation every clock cycle	Evaluation at minimum time-step or greater
Synchronous Designs only	Synchronous and Asynchronous Designs
Behavioural Simulation only	Behavioural, Functional and Timing Simulations
Faster Simulation Speed	Slower Simulation Speed

Table 2.
A comparison between cycle-driven simulation and event-driven simulation.

One advantage of CPLD devices is that they use non-volatile memory to store their configuration. As a result, their logic is already configured at power-up. This makes them ideal devices for systems where the logic is required to be ready for initialisation, such as glue logic for circuits.

Figure 1 shows the internal structure of a CPLD. These logic blocks consist of programmable PAL blocks. The inputs can be connected together to different AND gates using programmable fuses. The OR gate connections are fixed and cannot be reconfigured. Although less configurable than a PLA (which contains both programmable AND and OR planes), this reduction in complexity makes PAL blocks cheaper to manufacture. In order for PAL blocks to be able to implement sequential designs, a D flip-flop can be used to store the state of the output. CPLDs can connect multiple logic blocks together using the programmable interconnection matrix in order to implement more complex designs.

While CPLD devices are still used for specific tasks, the most common PLD in use today is the Field Programmable Gate Array (FPGA). These devices are quite similar in structure to the mask-programmed gate array (MPGA) [10] which was one of the first commercial programmable PLDs available. One benefit of using FPGAs is that they can be electronically reconfigured, whereas the previous MPGAs configuration was specified at the time of manufacture. The first FPGA, the Xilinx XC2064 was invented by Ross Freeman and Bernard Vonderschmitt in 1985 [11]. Early FPGAs were mainly used in the telecommunications and networking sectors as they were often cheaper than manufacturing custom silicon for these tasks.

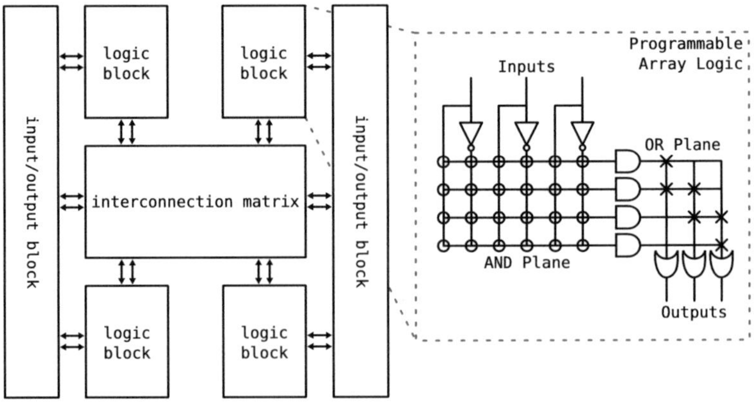

Figure 1.
Structure of a Complex Programmable Logic Device (CPLD) and Programmable Array Logic (PAL) block. The programmable AND plane and the fixed OR plane are shown on the right.

Figure 2 shows the internal structure of the FPGA. Similarities can be seen between FPGA and CPLD devices where a programmable interconnect is used to connect programmable logic blocks. In an FPGA, the Configurable Logic Blocks (CLB) consist of Look Up Tables (LUTs). The output of these programmable memories are defined by their input signals. The multiplexer then selects either the output of the LUT or the D flip flop to allow for combinational or sequential logic, similar to PAL blocks in CPLDs. These blocks can then be connected together using the Switching Blocks (SB). Modern FPGAs are able to replicate the behaviour of tens of millions of logic gates and contain logic like RAM and multipliers. Today, they are often used in high-performance computing applications due to their performance and efficiency over processor-based algorithms.

2.3 Digital design flow

In digital design, it is often not practical to use gate-level descriptions. Instead, a representation called Register Transfer Level (RTL) is used. RTL allows for a high-level model of the design to be represented without having to think about the low-level logic structures required to implement the functionality [12]. This abstraction uses constructs like logic statements, arithmetic operations and control flow. Similarities can be drawn between programming using mnemonics in Assembly Language compared to functional programming in C. Using RTL allows the designer to focus on the functionality of the design rather than on the implementation. **Figure 3** presents the different stages a digital design must goes through in order to convert a RTL representation into an implementable design.

When a high level language is used for programming, the code written by the programmer must first go through a process called compilation before the code is executed. Similarly with digital hardware, a design specified using RTL (often using a HDL like VHDL or SystemVerilog) must go through a process called logic synthesis. This process analyses the given RTL and converts it into a set of primitives that is functionally equivalent. Primitives are the basic building blocks of any logic design and consists of both combinational and sequential blocks. Examples include boolean logic (NOT/AND/OR/XOR etc.), multiplexers and flip-flops. This output is stored in a file called a net-list, which contains a list of all the primitive blocks and the nets that connect them together.

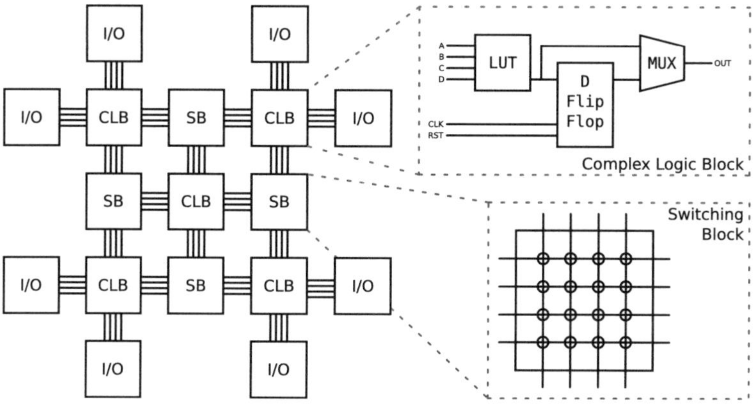

Figure 2.
Structure of a Field Programmable Gate Array (FPGA). The Complex Logic Block (CLB) consists of a programmable memory called a Look Up Table (LUT), a D flip flop to store state and a multiplexer to select the output signal. The programmable Switching Block (SB) is used to connect the CLBs together.

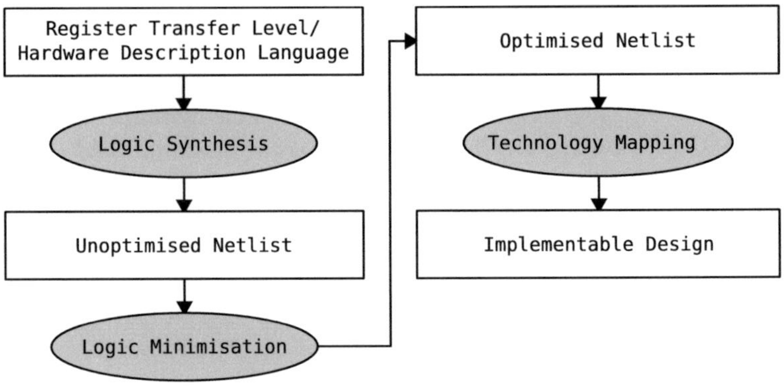

Figure 3.
A flowchart showing the different stages of the logic synthesis/digital design process.

The net-list generated from the logic synthesis is not optimized and must go through a process known as logic minimization. There can be many parameters to optimize for, such as area usage, power consumption and timing delay. There are many different methods that can be used to perform this optimization. Some early algorithmic methods include Karnaugh maps [13] and the Quine–McCluskey algorithm [14]. However as designs have become increasing complex, these algorithmic methods are not computationally feasible. This has lead to the use of heuristic optimizers such as ESPRESSO [15] and BOOM [16]. When using heuristic optimizers, it cannot be guaranteed that the minimized design is the global minimum. However in practice, these methods are sufficient and are widely used in logic synthesis tools today.

Following the logic minimization process, the optimized net-list is in an intermediate representation. A process called Technology Mapping must be performed before the design can be implemented in silicon or on a PLD. For silicon, this intermediate representation is compared against a library of available "building blocks" called leaf-level cells. The mapper then selects and connects these leaf-level cells, rebuilding the circuit. Further optimization may be performed here as the available leaf cells may be able to replace multiple blocks in the intermediate representation. For PLDs, the process is similar. The PAL blocks in CPLDs and the LUTs in FPGAs can be configured to replace one or multiple blocks. These are then connected together using the programmable interconnects. In comparison to logic minimization, the optimizations performed here are much simpler. After the technology mapping process is complete, the designer now has an implementable design. This is often in the GDSII/OASIS format for silicon manufacturing and in a bitstream format for PLDs. The top EDA companies for ASIC digital design tools include Cadence Design Systems, Siemens and Synopsys, with Xilinx and Intel providing FPGA tools.

2.4 Grammatical evolution and circuit design

Grammatical Evolution (GE), the tool used in this chapter and described in detail in the next section, has been used to evolve Verilog circuits, such as the one-bit adder and D-type latch at the gate level [17]. Notably, the one-bit adder is frequently used as a case study to evolve combinational circuits at the gate-level through GE [18–20]. However, gate-level evolution is less likely to scale to highly complex circuits from scratch [21]. In response to scalability issues inherent in gate level evolution, [22] proposed functional level evolution through Genetic Algorithms, which uses higher-level functions such as multiplexers, adders, subtractors instead of primitive gates to help reduce the search space. Similarly, [23] evolved a 3-bit multiplier using only binary multiplexers. 9- and 25-Median approximate circuits have also been designed at the functional level through Cartesian GP [24]. We address the scalability concern by performing circuit evolution through GE at a more abstract level – RTL modeling, where the focus is on describing the circuit's behavior [25, 26].

3. Grammatical evolution

Biological evolution has been a source of inspiration for many techniques that formed the field of evolutionary computation (EC), and has been used to address a wide range of problem domains ranging from the small to the huge, solving molecular to astronomical related problems. One of the most successful evolutionary techniques is GP, introduced by John R. Koza in his book "Genetic Programming—On the programming of Computers by Means of Natural Selection" [27], which mimics natural selection in an iterative way to find an optimal (best) solution. Algorithm 1 details the steps required to implement a standard GP. A survey of the different GP techniques current available in the literature is out of the scope of this work, the interested reader can find in [28] a comprehensive review of various aspects and techniques of GP and their categorization.

Algorithm 1: Standard GP pseudocode

Input: GP parameters setup.
Output: Best Individual (solution).
PopSize = No. of individuals
MaxGen = No. of generations (stop criterium)
Pop ← CreateIndividuals(PopSize)
while *a stop condition is not met <MaxGen>:* **do**
    Pop ← EvaluatePop(Pop)
    Parents ← SelectFittest(Pop)
    Offspring ← Crossover(Parents)
    Offspring ← Mutation(Offspring)
    Replace(Offspring, Pop)

Grammatical evolution (GE) is an evolutionary computation and, more specifically, a genetic programming (GP) technique [29] that addresses the closure issue of Koza-style GP, which effectively confines GP to single-type problems. This is achieved through the use of a grammar, generally in Backus-Naur Form (BNF) [29, 30], or Attribute Grammar (AG) [31–34].

The GE system shown in **Figure 4** automatically generates programs using three main components: (i) grammar; (ii) cost function; and (iii) search engine. The grammar describes the program's syntax, the cost function evaluates the quality of each program, and the search engine, typically a GA, searches within the program space defined by the grammar.

In GE, a typical representation for an individual is a binary string grouped into codons (e.g. 8 bits). The linear representation of the genome allows the application of genetic operators such as crossover and mutation in the manner of a typical GA, unlike tree-based GP.

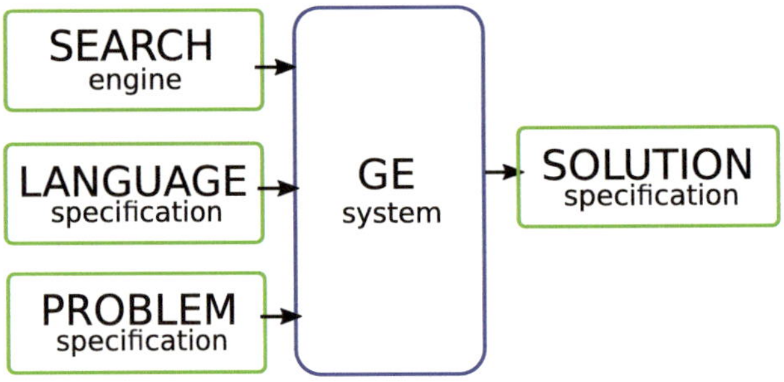

Figure 4.
The GE system uses a search engine (typically a GA) to generate solutions for a given problem, by recombining the genetic material (genotype) and mapped onto programs (phenotype) according to a language specification (interpreter/compiler).

3.1 Initialisation

In GP, the standard initialisation is the ramped-half-and-half (RHH) technique, introduced in [27]. In order to ensure diversity in the population, GP individuals typically represented as trees are created with different depths. The RHH technique uses two methods to create a tree: full and grow. Typically there is a probability of 0.5 to select either method for a particular individual. The full method creates trees with full branches at the maximum specified depth, whereas the grow method creates trees with different length of branches and different depth size up to the maximum allowed.

Sensible Initialisation is an adaptation of ramped-half-and-half initialisation routine in GP [35]. SI requires the grammar to labelled— whether a production rule is recursive or non-recursive and the minimum derivation tree depth to fully expand a production rule. SI requires a maximum tree depth to be specified prior. Similar to ramped-half-and-half in GP, SI applies both the grow and full method. When applying the grow method any production can be selected while the full method chooses only recursive productions. Both grow and full methods are subject to the constraints of not exceeding the maximum specified tree depth and the availability of enough tree depth budget to fully expand all non-terminals to terminals.

3.2 GE operators

Generally, GE uses a one-point crossover as it has been shown to be effective [36]. In crossover, two individuals are selected as parents and a single crossover point within each parent's genome is randomly chosen, dividing the genome into two halves: left and right sub-genomes. The right sub-genomes of both parents are swapped to create two offspring. However, crossover points that lie within non-coding regions (unused codon(s) from the mapping step) may not be so useful. As a result, a variant of one-point crossover known as *effective crossover*, which constrains the selection of the crossover point to be within the effective length of an individual's genome is preferred. Point mutation is a commonly used mutation operator in GE. Each bit within the binary string genome is mutated or flipped using the specified mutation probability. However, neutral mutations can occur whereby mutated codons select the same productions as the original codon; as a result, the corresponding phenotype remains the same.

3.3 GE example

To illustrate the application of GE, we first explain the evolutionary process using a mathematical optimisation problem as study case.

GE begins with the start symbol of the grammar, then the codons are used to select and apply the grammar production rules to finally build a program. This mapping process is illustrated in **Figure 5** with a simple example, where the production rules in the grammar contains a set of user-defined functions: $max\ (a, b)$, $min\ (a, b)$, $addition(a, b)$, $subtraction(a, b)$, $multiplication(a, b)$, $division(a, b)$, const, and X, which is a value (or a vector) sampled from the independent variable.

The production rules for each non-terminal are indexed starting from 0 and, when selecting a production rule (starting with the left-most non-terminal of the developing program) the next codon value in the genome is read and interpreted using the formula: $p = c \ \% \ r$, where c represents the current codon value, % represents the modulus operator, and r is the number of production rules for the left-most non-terminal.

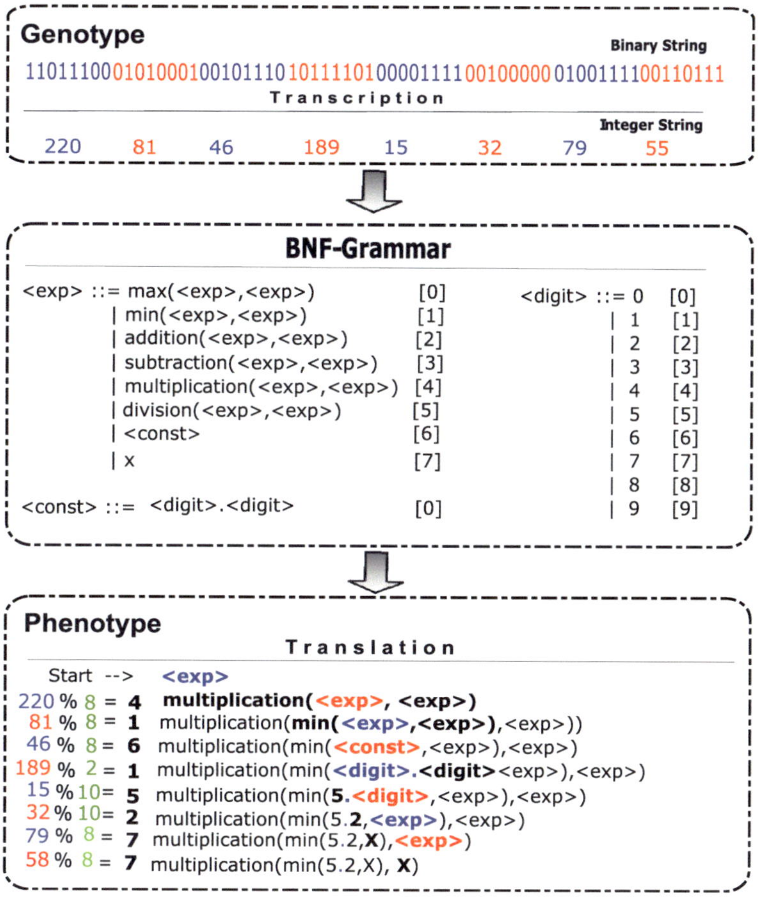

Figure 5.
Example of a GE genotype-phenotype mapping process for the Iris dataset, where the binary genotype is grouped into codons (e.g. 8 bits; red & blue), transcribed into an integer string, then used to select production rules from a predefined grammar (BNF-Grammar), and finally translated into a sequence of rules to build a solution (phenotype).

To prevent reaching the end of the genome without consuming all the available codons, then a wrapping process is used to continuing reading from the beginning of the genome. This mapping process stops when all of the non-terminal symbols have been replaced, in order to get a valid program. An exemption to this process is in the case when it fails to replace all of the non-terminal symbols after a maximum number of iterations, then it is considered an invalid individual and it is penalized with the lowest possible fitness.

4. ADDC

ADDC is an evolutionary HDL circuit design framework mainly driven by GE. ADDC requires a grammar and a testbench as inputs for circuit evolution and verification respectively. The designed grammar must be BNF compliant and must satisfy the grammar sufficiency property. Thus, the grammar must contain all the necessary building blocks required to potentially evolve an optimal circuit. ADDC is technology agnostic and easily configurable as the choice of HDL and simulator are left to the user to choose. Illustrated in **Figure 6** is ADDC's design flow for functional evolution of circuits.

During the initial phase of the circuit design process, ADDC creates an initial population of circuit designs using a suitable GE initialisation routine such as sensible

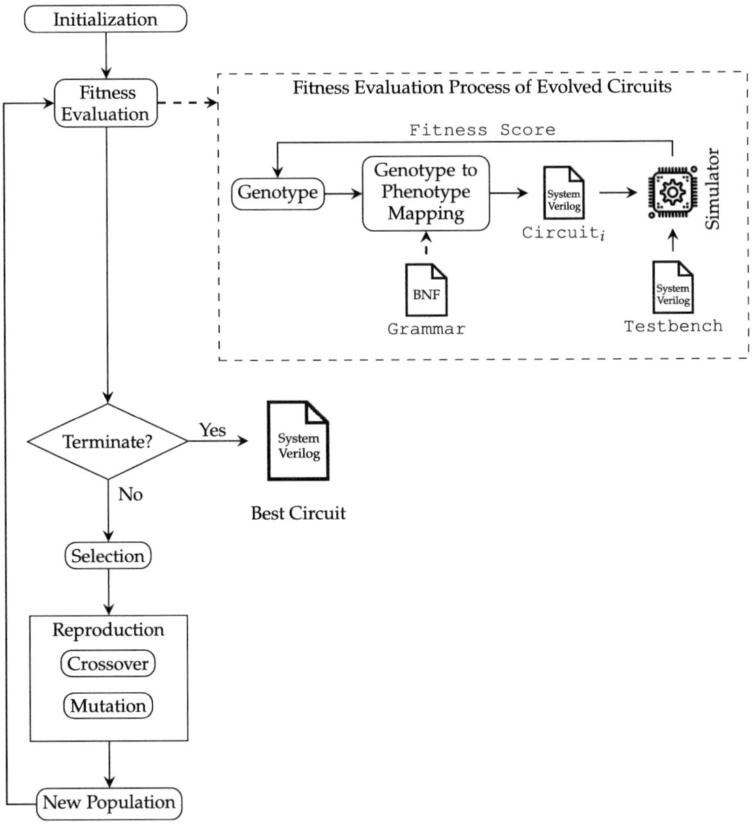

Figure 6.
ADDC Functional Circuit Evolution Overview.

initialisation. These individuals then undergo fitness evaluation. The fitness evaluation phase entails a number of steps. First, the genotype (genome) of each individual is translated to a HDL (SystemVerilog in this work) circuit design (phenotype) by the GE mapper, using the grammar designed for the circuit. Functional simulation of each circuit takes place, assuming all circuit designs are valid. For these experiments, *Icarus Verilog*, a lightweight and open-source Verilog simulator is used. The simulator uses a testbench which must be provided by the user for circuit verification. A testbench is a verification description written in the same HDL as the circuit that ensures the device under test (DUT) or evolved circuit meets functional and timing requirements. A typical testbench uses a test vector(s) which contains circuit inputs and their respective expected outputs. For example, test vector(s) for combinational circuits are usually created from their truth table. The simulator drives these input(s) through the DUT and verifies if the circuit output is the desired output. The sum of the number of passed cases out of the total number of cases represents the fitness value of the circuit. After all circuits have been evaluated, ADDC makes available the best circuit design if the specified termination criterion is satisfied, otherwise the circuit design evolution continues.

The next phase is reproduction, where usually individuals with either good overall fitness score or individuals that perform best on certain cases are selected for crossover and mutation to create a new population of circuit designs. Lexicase selection performs well on circuit design benchmarks [25, 26], hence selected as the choice of selection routine. Also, depending on the genetic algorithm (GA) of choice, for example steady state, generational GAs etc., events like replacement or elitism may take place in creating the new population. The new population undergoes fitness evaluation in similar manner as described in the previous section. The process continues until the termination criterion is satisfied and the best circuit design returned as solution.

5. Experiments

Three circuit benchmark problems are considered, namely: *Hamming Code (7,4) Encoder* (corresponding decoder evolved in [25]), *Seven Segment Display* [25, 37] and *16-to-4 Multiplexer* [25, 38, 39]. These problems are not only standard benchmarks in circuit design literature, but are used in real world applications. For example, hamming codes, seven segment displays and multiplexers are used in satellite communication, digital calculators and telephone networks respectively.

5.1 Benchmark problems

5.1.1 Hamming code (7,4) encoder

Hamming codes are a linear error-correcting codes capable of detecting a single error and at most two errors, but are only capable of correcting a single error. They belong to a category of codes referred to as Linear Block Codes. A Hamming Code (7,4) Encoder encodes a 4-bit data word into a 7-bit code word prior to data transmission by generating and adding three parity bits to the data word.

The structure of the code words generated by hamming codes can be classified into two categories: *systematic* and *non-systematic* encodings. The structure of systematic encoding separates the data word and code word while the data word and parity bits are interspersed in non-systematic encoding. We adopt systematic encoding as it is easier to separate the data word from the parity bits.

The grammar designed for evolving the Hamming Code (7,4) Encoder is shown in **Figure** 7. The circuit's interface is defined using the ⟨begin-module⟩ rule. The grammar uses four bitwise operators and a bitwise negation operator in Verilog defined using ⟨bitwise-op⟩ and ⟨bitwise-neg⟩ rules respectively. Hamming Code (7,4) Encoder generates three redundant bits stated earlier and these have been defined using ⟨parity-bit-1⟩, ⟨parity-bit-2⟩ and ⟨parity-bit-3⟩ rules. The *expr* rule is use by the ⟨parity-bit- $*$ ⟩ rules to evolve variable length expressions that generate the correct parity bit. The grammar also features an important Verilog construct, the `always` procedural block, defined using ⟨always-block⟩ which behaves similarly to an infinite loop and which is triggered by a change in any signal (indicated with $*$) to evaluate the statements within its scope.

5.1.2 Seven segment display

A Seven Segment Display is an electronic device used for the display of decimal numerals. It is also capable of displaying letters, though some letters such as K, X, Z etc. are difficult to recognise on the device. The specification for Seven Segment Display considered in this work supports decimal numerals (0-9) and A-F letter representations. These numbers and characters are encoded as a 4-bit binary number (0000–1111) referred to as binary coded decimal (BCD) which are sent as inputs to the Seven Segment Display. The 4-bit binary numbers starting from 0000 to 1001 are used to encode decimal numbers 0 to 9 respectively; while 1010 to 1111 are used to encode letters A-F. Upon receipt, the Seven Segment generates a 7-bit binary number (each bit corresponding an ON/OFF state of a segment) to turn on the appropriate LEDs to

```
⟨design-module⟩  ::=  ⟨begin-module⟩⟨code-block⟩⟨end-module⟩
   ⟨code-block⟩  ::=  ⟨declarations⟩⟨always-block⟩
  ⟨declaration⟩  ::=  "logic [1:3] parity_bit; "⟨nextline⟩
 ⟨always-block⟩  ::=  always@(*) begin ⟨nextline⟩⟨parity-bit-1⟩⟨parity-bit-2⟩
                      ⟨parity-bit-3⟩⟨output-stmt⟩end⟨nextline⟩
 ⟨parity-bit-1⟩  ::=  parity_bit[1] "=" ⟨expr⟩;⟨nextline⟩
 ⟨parity-bit-2⟩  ::=  parity_bit[2] "=" ⟨expr⟩;⟨nextline⟩
 ⟨parity-bit-3⟩  ::=  parity_bit[3] "=" ⟨expr⟩;⟨nextline⟩
  ⟨output-stmt⟩  ::=  codeword "=" {dataword,parity_bit}⟨nextline⟩
         ⟨expr⟩  ::=  ⟨input⟩ | ⟨bitwise-neg⟩⟨input⟩ | ⟨input⟩⟨bitwise-op⟩⟨input⟩
                      | ⟨input⟩⟨bitwise-op⟩⟨expr⟩ | ⟨bitwise-neg⟩(⟨expr⟩)
 ⟨dataword-idx⟩  ::=  1 | 2 | 3 | 4
   ⟨bitwise-op⟩  ::=  & | "|" | ^ | ^~
  ⟨bitwise-neg⟩  ::=  ~
        ⟨input⟩  ::=  dataword[⟨dataword-index⟩]
   ⟨codewd-bit⟩  ::=  ⟨input⟩ | parity_bit[1] | parity_bit[2] | parity_bit[3]
 ⟨begin-module⟩  ::=  "module hmc_encoder(input [1:4] dataword, output logic
                      [1:7] codeword);" ⟨nextline⟩
   ⟨end-module⟩  ::=  endmodule
     ⟨nextline⟩  ::=  "\n"
```

Figure 7.
Hamming code (7,4) encoder grammar.

display the digit/letter. **Figure 8** shows the grammar designed to evolve the seven segment display. The BCD and the 7-bit binary numbers are defined by the ⟨bcd⟩ and ⟨seven-segment⟩ rules respectively. The grammar uses `switch-case` construct defined using the ⟨switch-case⟩ as well as the always procedural block (⟨always⟩). ⟨begin – module⟩ defines the circuit interface of the Seven Segment Display.

5.1.3 16-to-4 multiplexer

A multiplexer is a multiple-input single-output device that accepts data (data lines) and an address (select lines) as inputs and uses the address to select the corresponding data line to be transmitted. The 16-to-4 multiplexer has 16 data lines and 4 select lines.

Figure 9 shows the grammar designed to evolve the multiplexer. Similar to the Seven Segment Display Grammar, the 16-to-4 Multiplexer Grammar also uses the always procedural block. However, here an `if-else` (⟨if-else⟩) construct is used, although the `switch-case` construct is also suitable in this context. The addresses used to select a data line as output are defined using the ⟨address⟩ rule. The ⟨address⟩ is used to generate a conditional statement (⟨cond⟩) by the ⟨if-else⟩ rule to determine the data line to select. The ⟨data-index⟩ defines the indexes of the 16-bit data which is used by the ⟨data-bit rule⟩ to select data line. ⟨begin-module⟩ defines the circuit interface of the 16-to-4 Multiplexer.

5.2 Evolutionary parameters

Experimental parameters used for running the experiments are shown in **Table 3**. The generation number and population size were selected based on preliminary

```
 ⟨design-module⟩ ::= ⟨begin-module⟩⟨next-line⟩⟨code-block⟩⟨end-module⟩
    ⟨code-block⟩ ::= ⟨always⟩
        ⟨always⟩ ::= always@(bcd)⟨nextline⟩begin⟨nextline⟩⟨switch-case⟩⟨nextline⟩
                     end
   ⟨switch-case⟩ ::= case(bcd) ⟨next-line⟩⟨case-stmt⟩⟨nextline⟩ endcase
     ⟨case-stmt⟩ ::= ⟨bcd⟩ ":"⟨output⟩"="⟨seven-segment⟩;
                   | ⟨bcd⟩ ":"⟨output⟩"="⟨seven-segment⟩;⟨nextline⟩⟨case-stmt⟩
           ⟨bcd⟩ ::= 4'b0000 | 4'b0001 | 4'b0010 | 4'b0011 | 4'b0100
                   | 4'b0101 | 4'b0110 | 4'b0111 | 4'b1000 | 4'b1001
                   | 4'b1010 | 4'b1011 | 4'b1100 | 4'b1101 | 4'b1110
                   | 4'b1111
 ⟨seven-segment⟩ ::= 7'b1111110 | 7'b0110000 | 7'b1101101 | 7'b1111001
                   | 7'b0110011 | 7'b1011011 | 7'b1011111 | 7'b1110000
                   | 7'b1111111 | 7'b1111011 | 7'b1110111 | 7'b0011111
                   | 7'b1001110 | 7'b0111101 | 7'b1001111 | 7'b1000111
        ⟨output⟩ ::= segment
  ⟨begin-module⟩ ::= "module seven_segment_display(output logic[6:0]
                     segment, input logic[3:0] bcd);"
    ⟨end-module⟩ ::= endmodule
      ⟨nextline⟩ ::= "\n"
```

Figure 8.
Seven segment display grammar.

```
⟨design-module⟩ ::= ⟨begin-module⟩⟨nextline⟩⟨code-block⟩⟨end-module⟩
   ⟨code-block⟩ ::= ⟨always⟩
       ⟨always⟩ ::= always@(*)⟨nextline⟩⟨stmt⟩
         ⟨stmt⟩ ::= ⟨if-else⟩⟨nextline⟩
      ⟨if-else⟩ ::= if(⟨cond⟩)⟨expr⟩⟨nextline⟩else⟨expr⟩
                  | if(⟨cond⟩)⟨expr⟩⟨nextline⟩⟨else-if⟩else⟨expr⟩
      ⟨else-if⟩ ::= else if(⟨cond⟩)⟨expr⟩⟨nextline⟩
                  | else if(⟨cond⟩)⟨expr⟩⟨nextline⟩⟨else-if⟩
         ⟨cond⟩ ::= address "==" ⟨address⟩
         ⟨expr⟩ ::= out "=" ⟨data-bit⟩
      ⟨address⟩ ::= 4'b0000 | 4'b0001 | 4'b0010 | 4'b0011 | 4'b0100 | 4'b0101
                  | 4'b0110 | 4'b0111 | 4'b1000 | 4'b1001 | 4'b1010 | 4'b1011
                  | 4'b1100 | 4'b1101 | 4'b1110 | 4'b1111
   ⟨data-index⟩ ::= 1 | 2 | 3 | 4 | 5 | 6 | 7 | 8 | 9 | 10 | 11 | 12 | 13
                  | 14 | 15 | 16
     ⟨data-bit⟩ ::= data[⟨data-index⟩]
 ⟨begin-module⟩ ::= "module mux(output logic out, input logic [1:4] address,
                    input logic [1:16] data);"
   ⟨end-module⟩ ::= endmodule
     ⟨nextline⟩ ::= "\n"
```

Figure 9.
16-to-4 multiplexer grammar.

Parameter	Value
Initialization	Sensible Initialization
No of generations	50 *for Hamming Code (7,4) Encoder for Seven Segment Display* & *16-to-4 Multiplexer*
Mutation rate	0.01
Crossover rate	0.8
Replacement rate	0.5
No of independent runs	50
Population	1,000
Selection	Lexicase Parent Selection

Table 3.
Experimental run parameters.

experiments. The generation sizes used for the preliminary experiments were 50, 100 and 200; the population sizes were 100, 200, 500, 1000 and 2000. For each problem, 5 independent runs were conducted. The choice of generation number and population size for the actual experiments were based on setups with majority of the runs with mean best fitness of the final generation within the fourth quartile of the maximum fitness. The other parameters used remain the same as used in [25, 26].

50 generations were used for evolving the Hamming Code (7,4) Encoder, while 100 generations was used for each of the Seven Segment Display and 16-to-4 Multiplexer designs as preliminary results revealed these problems were relatively

challenging to evolve compared to the Hamming Code (7,4) Encoder. All other parameters remain the same for all benchmark problems.

5.3 Training and testing

The number of training and testing cases are tabulated in **Table 4**.

Each of the Hamming Code (7,4) Encoder and Seven Segment Display have only 16 cases. However, for the Hamming Code (7,4) Encoder every correct bit in each bit position in the codeword is counted as part of the total fitness score for a candidate circuit, giving a total of 112 (7 × 16) cases. Given that Hamming Code (7,4) and Seven Segment Display have so few cases, it is feasible to perform exhaustive testing during training, leaving no cases for testing as shown in **Table 4**.

On the other hand, the 16-to-4 Multiplexer has 2^{20} cases making exhaustive testing infeasible. As a result we uniformly sample 4,100 and 5,000 cases for training and testing respectively as shown in **Table 4**.

5.4 Results and discussions

Results obtained from experiments conducted demonstrate ADDC is ideal for evolving digital circuit designs due to the use GE and a HDL which permits designs to be done at a more abstract level. The evolutionary performance for the experiments conducted for Hamming Code (7,4) Encoder, Seven Segment Display and 16-to-4 Multiplexer described in Section 5.1 are visualized in **Figures 10–12** respectively. The success rate per benchmark problem is tabulated in **Table 5**. A representative solution per each circuit benchmark problem is shown in **Figures 13–15** in the Appendix. The advantages and disadvantages of the proposed approach is discussed in Section 5.7.

5.5 Success rate

A successful run is a single independent evolutionary run that evolved an optimal circuit for the target problem. Fifty independent runs were conducted for all three benchmark problems. The success rate is the number of successful runs divided by the total number of evolutionary runs (i.e. 50) as tabulated in **Table 5**. A 100% success rate was attained for the Hamming Code (7,4) Encoder. The Seven Segment Display and 16-to-4 Multiplexer obtained 60 and 86% success rates, respectively.

5.6 Evolutionary performance

Visualization of the evolutionary performance as evolution progressed for Hamming Code (7,4) Encoder, Seven Segment Display and 16-to-4 Multiplexer are shown in **Figures 10–12** respectively.

Benchmark problem	No of training cases	No of testing cases
Hamming Code (7,4) Encoder	112	—
Seven Segment Display (*with A-F letter representations*)	16	—
16-to-4 Multiplexer	4100	5000

Table 4.
Number of training and testing cases.

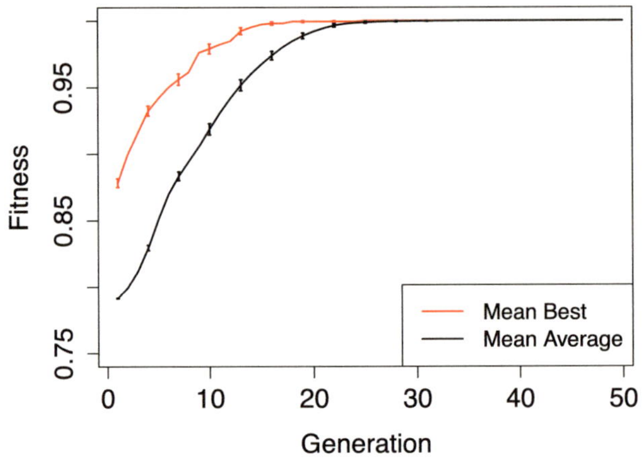

Figure 10.
Mean best and mean average across runs for hamming code (7,4) encoder.

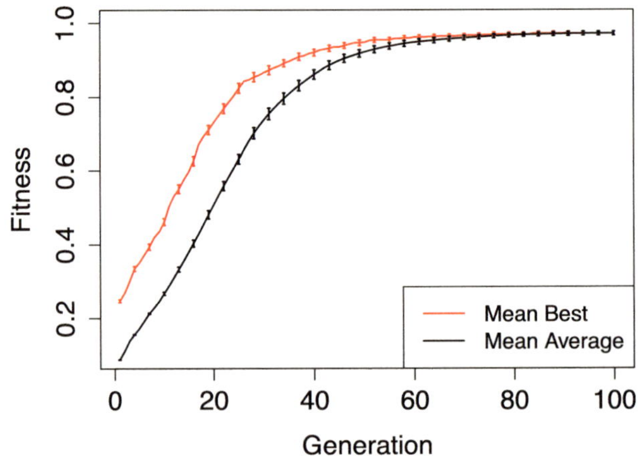

Figure 11.
Mean best and mean average across runs for seven segment display.

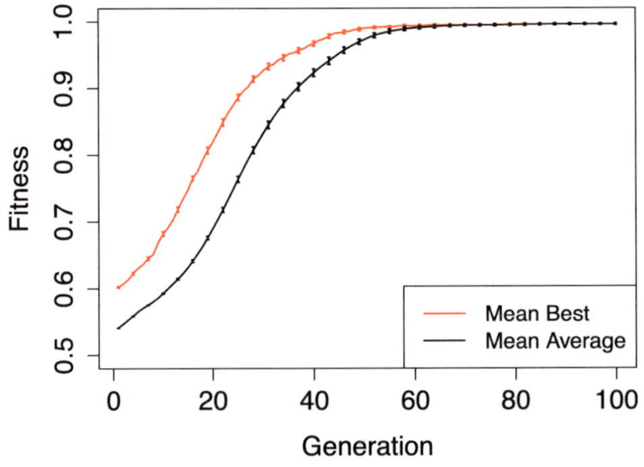

Figure 12.
Mean best and mean average across runs for 16-to-4 multiplexer.

Benchmark problem	Success rate
Hamming Code (7,4) Encoder	50/50
Seven Segment Display (*with A-F letter representations*)	30/50
16-to-4 Multiplexer	43/50

Table 5.
Success rate for benchmark problems.

The red line represents the mean best fitness per generations across the 50 independent runs conducted, while the black line represents the mean average fitness. Also plotted are error bars representing the standard error. The error bars are short to non-existent indicating small variability between the fitnesses of individuals. Furthermore, all three plots reveal a steady and progressive increase in fitness as the evolution progressed indicating the evolutionary search is continuously searching regions of the solutions where fitter individuals are located. Hamming Code (7,4) Encoder and 16-to-4 Multiplexer problems discover individual(s) that solve more that 50% of the test cases from the initial generations while the Seven Segment Display evolves individual(s) that solve 25% of the test cases on average.

5.7 Advantages and disadvantages of proposed approach

First, evolved circuit designs are quite interpretable compared to gate-level designs. This is due to the high level of abstraction at which these designs are performed which focuses on evolving circuit behaviours as opposed to evolving gate-level designs. Gate-level design approaches are challenging to scale to complex circuits [40]. The use of constructs such as `if-else`, `switch-case`, `always` procedural blocks, `bitwise` operators etc., makes it easier for humans to understand the behaviour of a circuit and if required effect manual modifications. Second, verification of circuits is easier as the circuit behaviour are easier to interpret enabling robust testbenches to be written.

Third, like any other methodology, there exist few disadvantages. The use of HDL requires the user to have technical knowledge about the HDL of choice— how to design grammars free of syntax errors and modelling errors. Syntax errors are easier to find and fix as most simulators will report such errors at the functional simulation phase. Modelling errors are a bit more challenging to fix, as they are only noticeable during synthesis (conversion of RTL or high level designs to gate-level representation) phase of the circuit design when designed grammars do not adhere to the guidelines of the HDL of choice. For example, fully functional representative solutions for the 16-to-4 Multiplexer and Seven Segment Display shown in **Figures 13** and **14** respectively may not be directly synthesizable (depending on the synthesis tool), as the `case` statements and `if` conditions are not mutually exclusive. Some of these errors can be fixed by defining new rules, modifying existing rules, the use of attribute grammars in order to impose the necessary constraints etc. The redundant logic can be gotten rid of by via a number of approaches: manually by hardware designer, design of SystemVerilog/Verilog redundant logic removal algorithm, use `unique case` statements (for switch case constructs) etc.

The choice of operators to use for evolving circuits is key as it has been shown to increase simulation time of circuits if inappropriate operators are chosen [26].

Furthermore, some circuit designs may contain redundant block of code which impede interpretability as observed in representative best solutions for 16-to-4 Multiplexer and Seven Segment Display shown in **Figures 13** and **14** respectively in Appendix. Only 16 of the if conditions and case statements are valid for the 16-to-4 Multiplexer and Seven Segment Display representation circuit designs respectively.

6. Conclusions and future directions

We have presented a system for the automated design of digital circuits, ADDC. ADDC is the next logical step in the evolution of Electronic Design Automation and this chapter has described how the history of integrated circuits has led to the confluence of GE, circuit simulators and HDLs. ADDC has been demonstrated on three difficult, real-world problems and was successful on all three of them, including one with 2^{20} possible inputs.

The HDLs employed here are hugely powerful and expressive. Digital designers often operate at very high levels of abstraction using *IP blocks*, which is somewhat similar to using libraries when programming software. IP blocks are typically very powerful and can have complexities equivalent to tens of thousands of gates. Making some of these blocks available to ADDC will dramatically increase its scalability and doing so will simply involve expanding the grammar.

As the problems scale up, the number of test cases can become astronomical, as was the case in this chapter. While in this case we randomly sampled the training and test cases, it is also possible to use a more intelligent approach. Recent work [41] has investigated using clustering to select a representative set of test cases. This will permit us to operate at greater scales with confidence.

A circuit that functions correctly on a simulator is not guaranteed to be fit for purpose when rendered in silicon. This is because there are often other considerations, such as silicon area, power dissipation and delay. Future work will use multi-objective optimisation to include pressure on individuals to adhere to these constraints too.

While some of our automatically generated circuits have successfully been implemented in silicon on a Xilinx Artix-7 FPGA, e.g., an 8-to-1 multiplexer [25], ADDC does not yet include that step in its toolchain; to be fully automated it will need to include this.

Acknowledgements

The authors are supported by Research Grants 13/RC/2094 and 16/IA/4605 from the Science Foundation Ireland and by Lero, the Irish Software Engineering Research Centre (www.lero.ie). The third is partially financed by the Coordenação de Aperfeiçoamento de Pessoal de Nível Superior—Brazil (CAPES), Finance Code 001, and Fundação de Amparo à Pesquisa do Estado do Rio de Janeiro (FAPERJ).

Conflict of interest

The authors declare no conflict of interest.

DOI: http://dx.doi.org/10.5772/intechopen.104410

Appendix

```
module mux(
    output logic out,
    input logic [1:4] addr,
    input logic [1:16] data
);
 always@(*)
     if(addr == 4'b1000) out = data[9];
     else if(addr == 4'b0100) out = data[5];
     else if(addr == 4'b0000) out = data[1];
     else if(addr == 4'b1101) out = data[14];
     else if(addr == 4'b1100) out = data[13];
     else if(addr == 4'b1011) out = data[12];
     else if(addr == 4'b0111) out = data[8];
     else if(addr == 4'b1010) out = data[11];
     else if(addr == 4'b0100) out = data[1];
     else if(addr == 4'b1000) out = data[3];
     else if(addr == 4'b1101) out = data[5];
     else if(addr == 4'b1100) out = data[13];
     else if(addr == 4'b1011) out = data[11];
     else if(addr == 4'b0011) out = data[4];
     else if(addr == 4'b1000) out = data[10];
     else if(addr == 4'b0011) out = data[12];
     else if(addr == 4'b0111) out = data[8];
     else if(addr == 4'b0100) out = data[13];
     else if(addr == 4'b1011) out = data[11];
     else if(addr == 4'b1000) out = data[9];
     else if(addr == 4'b0100) out = data[5];
     else if(addr == 4'b0000) out = data[9];
     else if(addr == 4'b0101) out = data[6];
     else if(addr == 4'b1100) out = data[13];
     else if(addr == 4'b1011) out = data[12];
     else if(addr == 4'b0111) out = data[8];
     else if(addr == 4'b1010) out = data[1];
     else if(addr == 4'b1100) out = data[12];
     else if(addr == 4'b1111) out = data[16];
     else if(addr == 4'b1001) out = data[10];
     else if(addr == 4'b0011) out = data[4];
     else if(addr == 4'b1101) out = data[14];
     else if(addr == 4'b1100) out = data[2];
     else if(addr == 4'b1101) out = data[3];
     else if(addr == 4'b1010) out = data[10];
     else if(addr == 4'b0010) out = data[3];
     else if(addr == 4'b1100) out = data[15];
     else if(addr == 4'b0001) out = data[2];
     else if(addr == 4'b1000) out = data[13];
     else if(addr == 4'b1011) out = data[4];
     else if(addr == 4'b0111) out = data[14];
     else if(addr == 4'b0110) out = data[7];
     else if(addr == 4'b1110) out = data[15];
     else if(addr == 4'b1111) out = data[6];
     else if(addr == 4'b1010) out = data[11];
     else if(addr == 4'b0101) out = data[13];
     else if(addr == 4'b1001) out = data[3];
     else out = data[9];
 endmodule
```

Figure 13.
16-to-4 multiplexer representative solution.

```
module  seven_segment_display(
    output logic[6:0] segment,
    input logic[3:0] bcd
);
always@(bcd)
 begin
 case(bcd)
     4'b0000 : segment = 7'b1111110;
     4'b1001 : segment = 7'b1111011;
     4'b0010 : segment = 7'b1101101;
     4'b0110 : segment = 7'b1011111;
     4'b1000 : segment = 7'b1111111;
     4'b0000 : segment = 7'b1111110;
     4'b1101 : segment = 7'b0111101;
     4'b1011 : segment = 7'b0011111;
     4'b0010 : segment = 7'b0011111;
     4'b0100 : segment = 7'b0110011;
     4'b1010 : segment = 7'b1110111;
     4'b0000 : segment = 7'b1111110;
     4'b1110 : segment = 7'b1001111;
     4'b1110 : segment = 7'b1000111;
     4'b0011 : segment = 7'b1111001;
     4'b0010 : segment = 7'b1001110;
     4'b0101 : segment = 7'b1011011;
     4'b0110 : segment = 7'b1011111;
     4'b1100 : segment = 7'b1001110;
     4'b1111 : segment = 7'b1000111;
     4'b1100 : segment = 7'b1001111;
     4'b0010 : segment = 7'b1101101;
     4'b1010 : segment = 7'b0110011;
     4'b0101 : segment = 7'b1011011;
     4'b0010 : segment = 7'b1011111;
     4'b0100 : segment = 7'b0110011;
     4'b1010 : segment = 7'b1110111;
     4'b0011 : segment = 7'b0110011;
     4'b0001 : segment = 7'b0110000;
     4'b1001 : segment = 7'b0111101;
     4'b1100 : segment = 7'b1111111;
     4'b0100 : segment = 7'b1101101;
     4'b0111 : segment = 7'b1110000;
     4'b1101 : segment = 7'b0111101;
     4'b0010 : segment = 7'b1000111;
     4'b0000 : segment = 7'b0011111;
     4'b0010 : segment = 7'b1101101;
     4'b1101 : segment = 7'b0111101;
     4'b1011 : segment = 7'b1001110;
     4'b1000 : segment = 7'b1011111;
     4'b0101 : segment = 7'b1111011;
     4'b1101 : segment = 7'b0110000;
     4'b1111 : segment = 7'b0110000;
 endcase
 end
endmodule
```

Figure 14.
Seven segment display representative solution.

```
module hmc_encoder(input [1:4] dataword, output logic [1:7]
    codeword);
 logic [1:3] parity_bit;
 always  @(*) begin
     parity_bit[1]  =  dataword[1] ^~ dataword[4] ^~ dataword[2];
     parity_bit[2]  =  dataword[3] ^~ dataword[1] ^~ dataword[4];
     parity_bit[3]  =  dataword[4] ^ dataword[2] ^ dataword[3];
     codeword  =  {dataword,parity_bit};
  end
 endmodule
```

Figure 15.
Hamming code (7,3) encoder representative solution.

Author details

Conor Ryan[1*†], Michael Tetteh[1†], Jack McEllin[1†], Douglas Mota-Dias[1,2†], Richard Conway[1†] and Enrique Naredo[1†]

1 University of Limerick, Limerick, Ireland

2 Rio de Janeiro State University, Rio de Janeiro, Brazil

*Address all correspondence to: conor.ryan@ul.ie

† These authors contributed equally.

References

[1] Conway T, Conway R, Mulvaney K, Mahony SO, Billon C, Khan MK, et al. Low power application specific processor for ISM band transceiver. In: IET Irish Signals and Systems Conference (ISSC 2010). Cork, Ireland: IET; 2010. pp. 272-277. Available from: https://digital-library.theiet.org/content/conferences/10.1049/cp.2010.0525

[2] Barbe DF. VHSIC Systems and Technology. Computer. 1981;**14**(2):13-22

[3] Dewey A. VHSIC Hardware Description (VHDL) Development Program. In: 20th Design Automation Conference Proceedings. Miami Beach (FL): IEEE Press; 1983. pp. 625-628

[4] Dewey A, Gadient A. VHDL Motivation. IEEE Design Test of Computers. 1986;**3**(2):12-16

[5] IEEE Standard VHDL Language Reference Manual. IEEE Std 1076-1987; 1988

[6] IEEE Standard Hardware Description Language Based on the Verilog(R) Hardware Description Language. IEEE Std 1364-1995. 1996

[7] Ap. COMPANY NEWS; Cadence to Buy Gateway Design. The New York Times. 1989. Available from: https://www.nytimes.com/1989/10/05/business/company-news-cadence-to-buy-gateway-design.html [Accessed: January 10, 2022]

[8] Verilog Hardware Description Language Reference Manual (LRM). Version 1.0. Open Verilog International (OVI); 1991

[9] IEEE Standard for SystemVerilog: Unified Hardware Design, Specification and Verification Language. IEEE Std 1800-2005. 2005

[10] Section 8—XC157. In: The Integrated Circuit Data Book. Motorola Semiconductor Products Inc.; 1968. p. 3

[11] Freeman R. Xilinx Inc. Configurable Electrical Circuit Having Configurable Logic Elements and Configurable Interconnects. US4870302; 1989

[12] De la Guia Solaz M, Conway R. Razor based programmable truncated multiply and accumulate, energy-reduction for efficient digital signal processing. IEEE Transactions on Very Large Scale Integration (VLSI) Systems. 2014;**23**(1): 189-193

[13] Karnaugh M. The Map Method for Synthesis of Combinational Logic Circuits. Transactions of the American Institute of Electrical Engineers, Part I: Communication and Electronics. 1953; **72**(5):593-599

[14] McCluskey EJ. Minimization of Boolean Functions. The Bell System Technical Journal. 1956;**35**(6):1417-1444

[15] Brayton R, Hachtel G, Hemachandra L, Newton A, Sangiovanni-Vincentelli A. A Comparison of Logic Minimization Strategies Using ESPRESSO: An APL Program Package for Partitioned Logic Minimization. In: Proceedings of the International Symposium on Circuits and Systems. New York (NY): IEEE Press; 1982. pp. 42-48

[16] Hlavicka J, Fiser P. BOOMa—A heuristic boolean minimizer. In: IEEE/ACM International Conference on Computer Aided Design. ICCAD 2001. IEEE/ACM Digest of Technical Papers (Cat. No.01CH37281). San Jose, CA, USA: IEEE; 2001. pp. 439-442

[17] Cullen J. Evolving Digital Circuits in an Industry Standard Hardware

Description Language. In: Li X, Kirley M, Zhang M, Green D, Ciesielski V, Abbass H, et al., editors. Simulated Evolution and Learning. Berlin, Heidelberg: Springer; 2008. pp. 514-523

[18] Youssef A, Majeed B, Ryan C. Optimizing combinational logic circuits using Grammatical Evolution. In: 2021 3rd Novel Intelligent and Leading Emerging Sciences Conference (NILES); 2021. pp. 87-92

[19] Karpuzcu UR. Automatic verilog code generation through grammatical evolution. In: Proceedings of the 7th Annual Workshop on Genetic and Evolutionary Computation. GECCO '05. New York, NY, USA: Association for Computing Machinery; 2005. pp. 394-397. DOI: 10.1145/1102256.1102346

[20] Kratochvil O, Osmera P, Popelka O. Parallel grammatical evolution for circuit optimization. In: Ao SI, Douglas C, Grundfest WS, Burgstone J, editors. Proceedings of the World Congress on Engineering and Computer Science, WCECS '09. vol. II. International Association of Engineers. San Francisco, USA: Newswood Limited; 2009. pp. 1032–1037. Available from: http://www.iaeng.org/publication/WCECS2009/WCECS2009_pp1032-1037.pdf

[21] Vassilev VK, Miller JE. Scalability problems of digital circuit evolution evolvability and efficient designs. In: Proceedings. The Second NASA/DoD Workshop on Evolvable Hardware; 2000. pp. 55-64

[22] Murakawa M, Yoshizawa S, Kajitani I, Furuya T, Iwata M, Higuchi T. Hardware evolution at function level. In: Voigt HM, Ebeling W, Rechenberg I, Schwefel HP, editors. Parallel Problem Solving from Nature — PPSN IV. Springer: Berlin, Heidelberg; 1996. pp. 62-71

[23] Vassilev VK, Miller JF. Embedding landscape neutrality to build a bridge from the conventional to a more efficient three-bit multiplier circuit. In: Proceedings of Genetic and Evolutionary Computation Conference. Morgan Kaufmann; 2000

[24] Vasicek Z, Sekanina L. Evolutionary approach to approximate digital circuits design. IEEE Transactions on Evolutionary Computation. 2015;**19**(3): 432-444

[25] Ryan C, Tetteh M, Dias D. Behavioural Modelling of Digital Circuits in System Verilog using Grammatical Evolution. In: Proceedings of the 12th International Joint Conference on Computational Intelligence—ECTA,. INSTICC. SciTePress; 2020. pp. 28-39

[26] Tetteh MK, Mota Dias D, Ryan C. Evolution of complex combinational logic circuits using grammatical evolution with systemverilog. In: Hu T, Lourenço N, Medvet E, editors. Genetic Programming. Cham: Springer International Publishing; 2021. pp. 146-161

[27] Koza JR. Genetic Programming—On the programming of Computers by Means of Natural Selection. Complex Adaptive Systems. Cambridge (MA): MIT Press; 1992

[28] Eiben SJ, Agoston E. From evolutionary computation to the evolution of things. Nature. 2015;**521** (7553):476-482. DOI: 10.1038/nature14544

[29] Ryan C, Collins JJ, O'Neill M. Grammatical evolution: Evolving programs for an arbitrary language. In: Banzhaf W, Poli R, Schoenauer M, Fogarty TC, editors. EuroGP. vol. 1391 of Lecture Notes in Computer Science. Berlin: Springer; 1998. pp. 83-96

[30] O'Neill M, Ryan C. Grammatical evolution. IEEE Trans Evolutionary Computation. 2001;**5**(4):349-358

[31] Patten JV, Ryan C. Attributed grammatical evolution using shared memory spaces and dynamically typed semantic function specification. In: Genetic Programming—18th European Conference, EuroGP 2015, Copenhagen, Denmark, April 8-10, 2015, Proceedings. 2015. pp. 105-112. DOI: 10.1007/978-3-319-16501-1_9

[32] Karim MR, Ryan C. On improving grammatical evolution performance in symbolic regression with attribute grammar. In: Genetic and Evolutionary Computation Conference, GECCO '14, Vancouver, BC, Canada, July 12-16, 2014, Companion Material Proceedings. 2014. pp. 139-140. DOI: 10.1145/2598394.2598488

[33] Karim MR, Ryan C. A new approach to solving 0-1 multiconstraint knapsack problems using attribute grammar with lookahead. In: Genetic Programming—14th European Conference, EuroGP 2011, Torino, Italy, April 27-29, 2011. Proceedings. 2011. pp. 250-261

[34] Karim MR, Ryan C. Degeneracy reduction or duplicate elimination? an analysis on the performance of attributed grammatical evolution with lookahead to solve the multiple knapsack problem. In: Nature Inspired Cooperative Strategies for Optimization, NICSO 2011, Cluj-Napoca, Romania, 2011. Vol. 387. Berlin: Springer. Studies in computational intelligence; 2011. pp. 247-266. DOI: 10.1007/978-3-642-24094-2_18

[35] Ryan C, Azad R. Sensible initialisation in grammatical evolution. In: GECCO 2003: Proceedings Of The Bird Of A Feather Workshops, Genetic And Evolutionary Computation Conference. Chigaco (IL): AAAI; 2003. pp. 142-145

[36] O'Neill M, Ryan C, Keijzer M, Cattolico M. Crossover in Grammatical Evolution. The Search Continues. Genetic Programming. 2001:337-347

[37] Miller J. Cartesian Genetic Programming. Cartesian Genetic Programming. 2011. pp. 17-34. DOI: 10.1007/978-3-642-17310-3_2

[38] Kruse R, Borgelt C, Klawonn F, Moewes C, Steinbrecher M, Held P. Fundamental evolutionary algorithms. Computational Intelligence: A Methodological Introduction. London (UK): Springer London; 2013. pp. 227-274. DOI: 10.1007/978-1-4471-5013-8_13

[39] Fredivianus N, Prothmann H, Schmeck H. XCS revisited: A novel discovery component for the extended classifier system. Simulated Evolution and Learning. Berlin, Heidelberg: Springer Berlin Heidelberg. Lecture Notes in Computer Science. 2010;**6457**: 289-298. DOI: 10.1007/978-3-642-17298-4_30

[40] Zdenek, V. Bridging the Gap Between Evolvable Hardware and Industry Using Cartesian Genetic Programming. 2018. DOI: 10.1007/978-3-319-67997-6_2

[41] Ryan C, Kshirsagar M, Gupt K, Rosenbauer L, Sullivan J. Hierarchical clustering driven test case selection in digital circuits. In: Proceedings of the 16th International Conference on Software Technologies—ICSOFT,. INSTICC. SciTePress; 2021. pp. 589-596

Chapter 5

Genetic Algorithms for Chemical Engineering Optimization Problems

Thi Anh-Nga Nguyen and Tuan-Anh Nguyen

Abstract

Chemical engineering processes are frequently composed of multiple complex phenomena. These systems can be represented by a set of several equations, which are referred to as mathematical model of the process. Optimization in chemical engineering utilizes specialized techniques to determine the values of the decision variables at which the performance of the process, measured as the objective function(s), is minimum or maximum. The profitability of the process improves remarkably as a result of this selection. This benefit has encouraged the broad application of optimization for important industrial challenges. However, many problems in chemical engineering processes are hard to find the optimum using gradient-based algorithms. For example, the cases when the objective functions of the processes are multimodal, discontinuous, or implicit. Genetic algorithms (GAs) are a kind of metaheuristic searching optimization methods, which are inspired by nature, the mechanics of natural evolution and genetics. Genetic algorithms have received significant attention due to their remarkable advantages over classical algorithms. Compared with traditional optimization approaches, GAs are straightforward, robust, capable of handling the non-differentiable, discontinuous, or multimodal problems. The purpose of this paper is to give several case studies using genetic algorithms in chemical engineering optimization problems.

Keywords: optimization, genetic algorithm, chemical engineering, modeling

1. Introduction

Chemical engineering processes are frequently composed of multiple complex phenomena. These systems can be represented by a set of several equations, such as $\mathbf{z} = \mathbf{f}(\mathbf{d}; \mathbf{x}; \mathbf{p})$, where $\mathbf{z}$ is the vector state of the system. The system of equations is referred to as the mathematical model of the process. The performance of the process is predicted by the model from the assigned data of several "input" variables, $\mathbf{d}$ and $\mathbf{x}$, and a group of parameters, $\mathbf{p}$. Among the input variables, some, referred as $\mathbf{x}$, can be changed and are known as design variables, while others, referred as $\mathbf{d}$, are predetermined. The performance of the process can be evaluated through a set of output variables, $\mathbf{y} = \mathbf{g}\ (\mathbf{z})$, referred as the function of state of the system. Optimization in chemical engineering utilizes specialized techniques to determine the values of the decision variables, **x**, at which the performance of the process, measured as the objective function(s), $I(\mathbf{x})$, is minimum or maximum. The profitability of the process improves remarkably as a result of this selection of input/operating/decision variables. This benefit has encouraged the broad application of optimization for important industrial challenges. However, many problems in chemical engineering processes are hard to find the optimum using gradient-based algorithms. For example, the cases when the objective functions of the processes are multimodal, discontinuous, or implicit. Genetic algorithms (GAs) are a kind of metaheuristic searching optimization methods, which are inspired by nature, the mechanics of natural evolution, and genetics [1]. Genetic algorithms have received significant attention due to their remarkable advantages over classical algorithms. Compared with traditional optimization approaches, GAs are straightforward, robust, capable of handling the non-differentiable, discontinuous, or multimodal problems. GAs have been effectively employed in a wide range of various engineering, manufacturing, and management applications [2]. The purpose of this chapter is to give several case studies using genetic algorithms in chemical engineering optimization problems. The case studies include the optimization of an autothermal ammonia synthesis reactor, a separation module using membrane technology, and data-driven modeling optimizations of solid oxide fuel cells.

2. Optimization of an autothermal ammonia synthesis reactor

In the chemical process industries, ammonia is one of the most widely manufactured inorganic compounds [3]. The majority of ammonia produced commercially is consumed in fertilizers, with the rest going into plastics, synthetic fibers and resins, pharmaceuticals, explosives, papers, and refrigeration [4]. As a result, modeling and optimization of ammonia synthesis process have received a significant attention from both the academia and industry. Ammonia is produced predominantly from the combination of elements such as nitrogen and hydrogen in a catalytic process using a promoted iron catalyst firstly established by Haber and Bosch as the reaction [4]:

$$N_2 + 3H_2 \rightleftharpoons 2NH_3 \tag{1}$$

The reaction is reversible and exothermic, releasing a significant amount of heat. In order to achieve a high conversion, the heat of the reaction should be removed. Therefore, the process is typically carried out in an autothermal synthesis reactor, in which the heat of reaction is utilized to preheat the feed gas and ensure the suitable temperature inside. The production of ammonia depends on several factors such as the reactor length, the operating pressure, temperature of the feed and reacted gas, the flow rate, and composition of the gas mixture. The optimization problem of the process is to maximize the economic return. Many studies discussing the modeling, simulation, and optimization of an autothermal ammonia synthesis reactor can be found in literature. Some of them can be mentioned here as in Babu et al. [5], Babu and Angira [6], Carvalho et al. [7], Edgar et al. [8], Ksasy et al. [9], Murase et al. [10], Upreti and Deb [11], Yusup et al. [12]. However, the model discussed in the studies of Edgar et al. [8], Murase et al. [10] has some minor errors and has been corrected in Upreti and Deb [11]. Moreover, the studies primarily focus on optimizing reactor length for a specific reactor top temperature, usually 694 K [6, 7, 12], or for a limited set of temperatures [9, 11]. However, as reported in some studies [11, 12], the economic return is determined by the top temperature and also the reactor length (the temperature of feed gas entering to the reaction zone). As a result, rather than a single variable problem of reactor length, the optimization problem should be viewed as a multivariable problem.

In the study [13], both the reactor length and the reactor top temperature are considered in the design variables for maximizing the profit return of the process. In order to solve the multivariate optimization problem, the cyclic coordinate search technique was employed. This method alters the value of one decision variable at a time, and for each coordinate direction, the golden section search was utilized to solve the single variable optimum problem. However, this traditional searching approach is prone to get caught in local optima. Therefore, the genetic algorithm has higher chance to obtain the global optimum profit of the process.

2.1 Problem formulation of ammonia synthesis reactor

The system discussed here is an autothermal synthesis reactor, which is described in [10] and contains the correction of the objective function reported in [6, 11]. The feed gas contains 21.75 mole% nitrogen, 65.25 mole% hydrogen, 5.0 mole% ammonia, 4.0 mole% methane, and 4.0 mole% argon. In an autothermal reactor, the feed gas mixture enters from the bottom of the reactor, flows upward, enters the catalyst zone from the top, and moves downward. In the catalyst zone, the reaction takes place at around 500°C and 200 atm of pressure. The heat generated by the reaction is utilized to preheat the feed gas mixture in counter current flow. **Figure 1** shows the schematic diagram of an autothermal ammonia synthesis reactor. The considered factors affecting the synthesis process are the temperature of feed gas at the entrance of the reaction zone (top temperature) and the reactor length. The goal of the optimal design is to determine the conditions that will give the highest economic return from the reactor operation.

2.1.1 Objective function

The return of the process, which is calculated from the value of the product gas (heating value and ammonia value), subtract the cost of feed gas (as a source of heat

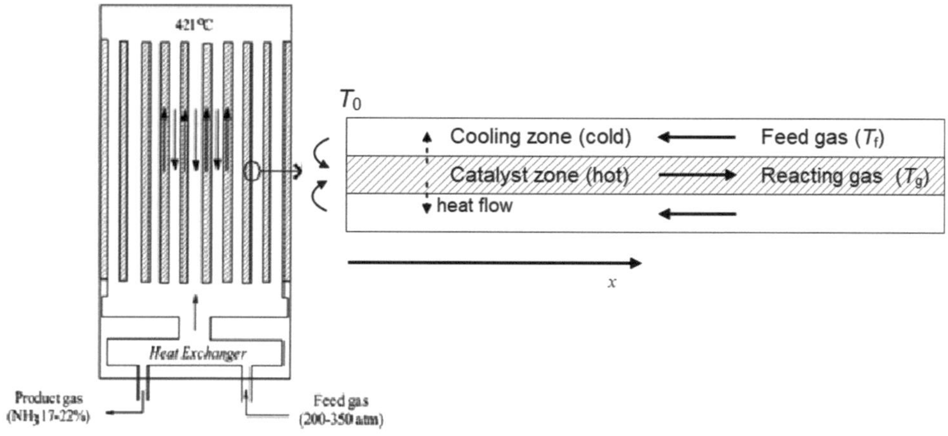

Figure 1.
Schematic diagram of an autothermal ammonia reactor [10].

only) and minus the amortization of reactor capital expenses, is the objective function for maximization (F). Other operating costs are not considered [11].

$$F = 1.3356 \times 10^7 - 1.708 \times 10^4 N_{N_2} + 704.09(T_g - T_0) - 699.27(T_f - T_0) \\ -(3.4566 \times 10^7 + 1.9837 \times 10^9 L)^{\frac{1}{2}} \tag{2}$$

in which,
T_f is the temperature of the feed gas.
T_g is the temperature of the reacting gas (the gas in the catalyst zone).
T_0 is the top temperature or temperature at the inlet of the catalyst zone.
N_{N_2} is the flow rate of nitrogen.
L is the length of the reactor.

2.1.2 Equality constraints

The heat balance for the feed gas and the reacting gas and the mass balance for the nitrogen flow along the catalyst zone, respectively, give the mathematical model for the system:

$$\frac{dT_f}{dx} = -\frac{US_1}{WC_{pf}}(T_g - T_f) \tag{3}$$

$$\frac{dT_g}{dx} = -\frac{US_1}{WC_{pg}}(T_g - T_f) + \frac{(-\Delta H)S_2}{WC_{pg}}\left(-\frac{dN_2}{dx}\right) \tag{4}$$

$$\frac{dN_{N_2}}{dx} = -f\left(k_1\frac{p_{N_2}p_{H_2}^{1.5}}{p_{NH_3}} - k_2\frac{p_{NH_3}}{p_{H_2}^{1.5}}\right) \tag{5}$$

in which

$$k_1 = 1.78954 \times 10^4 \exp\left(\frac{-20800}{RT_g}\right) \tag{6}$$

$$k_2 = 2.5714 \times 10^{16} \exp\left(\frac{-47400}{RT_g}\right) \tag{7}$$

$$p_{H_2} = 3p_{N_2} \tag{8}$$

$$p_{N_2} = \frac{286N_{N_2}}{2.598N_{N_2}^0 + 2N_N} \tag{9}$$

$$p_{NH_3} = \frac{286(2.23N_N^0 - 2N_{N_2})}{2.598N_N^0 + 2N_N} \tag{10}$$

The differential equations are valid in the interval $[0, L]$, in which L is the reactor length (or the length of the catalyst zone).

The notations T_f^0, T_g^0, and $N_{N_2}^0$ denote the initial value at $x = 0$ (at the inlet of the catalyst zone) for T_f, T_g, and N_{N_2}, respectively, and are given by

$$T_f^0 = T_g^0 = T_0, \quad N_{N_2}^0 = 701.2\text{kmol/hm}^2 \tag{11}$$

Other notations of the system are summarized in **Table 1**.

2.1.3 Box constraints

The variables are subjected to the following physical constraints, as is typical in industries [10]:

$$0 < L \le 10, \quad 400 \le T_f \le 800, \quad 0 \le N_{N_2} \le 3220 \tag{12}$$

The length of the reactor and the top temperature are chosen as the design variables. The remaining variables (T_f, T_g, and N_{N_2}) can be calculated from the model by three differential Eqs. (3), (4), and (5). Then, the objective function is determined by Eq. (2). Due to the constraints of temperatures, the variable T_0 is also set to be within 400 and 800.

The optimal design problem is summarized as follows:

$$\text{maximize } F = F(x, T_0)$$
$$s.t. \begin{cases} \dfrac{dT_f}{dx} = -\dfrac{US_1}{WC_{pf}}(T_g - T_f) \\ \dfrac{dT_f}{dx} = -\dfrac{US_1}{WC_{pg}}(T_g - T_f) + \dfrac{(-\Delta H)S_2}{WC_{pg}}\left(-\dfrac{dN_2}{dx}\right) \\ \dfrac{dN_2}{dx} = -f\left(k_1 \dfrac{p_{N_2} p_{H_2}^{1.5}}{p_{NH_3}} - k_2 \dfrac{p_{NH_3}}{p_{H_2}^{1.5}}\right) \\ T_f^0 = T_g^0 = T_0, \; N_{N_2}^0 = 701.2 \\ 400 \le T_f \le 800, \; 0 \le N_{N_2} \le 3220 \\ 0 < L \le 10, \; 400 \le T_0 \le 800 \end{cases} \tag{13}$$

Notation	
C_{pf}	Heat capacity of the feed gas
C_{pg}	Heat capacity of the reacting gas
f	Catalyst activity
ΔH	Heat of reaction
N	Mass flow of component designed by subscript
k	Reaction rate constant
p	Partial pressure of component designated by subscript
R	Universal gas constant
S_1	Surface area of catalyst tubes per unit length of reactor
S_2	Cross-sectional area of catalyst zone
U	Overall heat transfer coefficient
W	Total mass transfer flow rate

Table 1.
Notation of the synthesis system.

2.2 Optimization strategy

The system of ordinary differential Eqs. (3), (4), and (5) with initial conditions (11) was solved by Runge–Kutta fourth-order method. The system is well defined when the top temperature (T_0) and the interval of integration (the reactor length L) are assigned. After that, the economic return is clearly evaluated, and it can be considered as two-variable function of the top temperature and reactor length. The genetic algorithm is employed to find the optimal solution. In order to handle the constraints, a penalty or barrier is defined to the objective function whenever any of variable limits is violated. The proposed strategy is detailed as follows.

2.2.1 Genetic algorithm for optimization problem

The range of the design variables is $0 < x \leq 10, \; 400 \leq T_0 \leq 800$. The parameters of GA such as population size, crossover probability, mutation probability values were set to be 50, 1.0, and 0.30, respectively. The selection was chosen as roulette wheel selection with elitism. The number of generations was set to be 500.

2.2.2 Barrier method for constrained optimization

In barrier or penalty methods, the objective function will receive an undesired value when one of the constraints is violated. Therefore, the solution will be kept in the feasible region. The objective function has been modified as

$$F = \begin{cases} F \text{ if } 400 \leq T_f \leq 800, \; 0 \leq N_{N_2} \leq 3220 \\ 0 \text{ otherwise} \end{cases} \tag{14}$$

2.2.3 Parameters

The parameters were obtained from the literature [7, 10] and summarized in **Table 2**.

Parameter	Value	Unit
C_{pf}	0.707	kcal/kg K
C_{pg}	0.719	kcal/kg K
f	1.0	
ΔH	−26,000	kcal/kmol
R	1.987	kcal/kmol K
S_1	10	M
S_2	0.78	m^2
U	500	kcal/h m^2 K
W	26,400	kg/h

Table 2.
Model parameters [7].

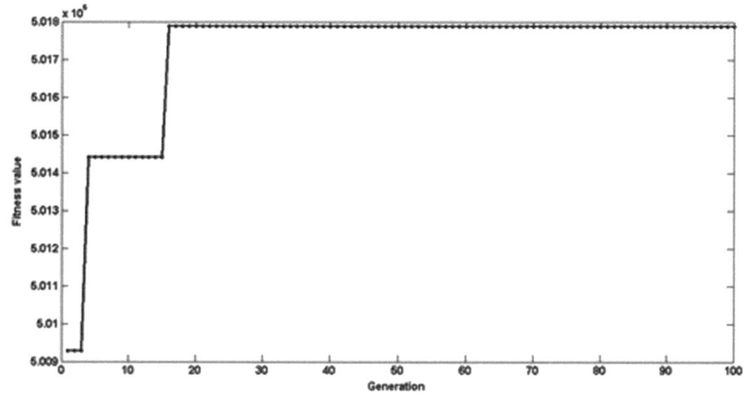

Figure 2.
Fitness value versus generations.

Variables	Interval	Cyclic coordinate [13]	Genetic algorithm
x (m)	[0,10]	6.724	6.772
T_0 (K)	[600,800]	700.27	707.09
T_f (K)	[400,800]	400.00	401.09
T_g (K)		629.94	631.12
N_{N2}	[0,3220]	490.68	490.68
F (10^6\$/year)		5.018	5.018

Table 3.
Maximization results.

2.3 Optimization results

Figure 2 shows the fitness values as a function of generation. As can be observed, the fitness function value achieved the highest after roughly 20 generations and then stayed unchanged. After 100 generations, it was obtained that the reactor length should be 6.772 m, and the top temperature should be 707.09 K. The process produces a profit of 5.018×10^6 \$ per year. The other parameters of the process are summarized in **Table 3** and compared with the findings of a cyclic coordinate search [13]. The profit value is slightly higher than those reported in the literature, which focused solely on reactor length optimization. From the results, the temperature at the entrance of the catalyst zone should be slightly higher, and that the reactor length should also be slightly longer than previously reported.

3. Economic optimization of membrane module for ultrafiltration of protein solution

The behavior of permeate flux has a significant impact on the performance of cross-flow ultrafiltration. Many factors cause flux declination, such as solution properties, membrane properties, and operation conditions. The majority of current research has centered on increasing membrane performance in terms of permeability and selectivity [14, 15]. Just a few studies have paid attention to the configuration and operation of the membrane module [16].

Various factors determine the decision of membrane module geometry for a given application, including fabrication method, power consumption, and fouling potential [17]. Manufacturers frequently recommend the membrane module design from the fabrication standpoint [17]. There is virtually no evidence that their approach prioritizes the energy efficiency. Currently, with a growing in energy concern and a falling in membrane cost, the membrane module design should place a higher attention on energy efficiency. As a result, it is necessary to propose a module design methodology that takes into account the energy factor.

Furthermore, membrane operating conditions are usually decided by user experience, a handbook, or a manual from the membrane supplier. However, the permeate flux equation governing the performance of the membrane system varies greatly between different situations. In this aspect, for any specific application, a general methodology for the design and operation conditions should be studied.

In cross-flow ultrafiltration of protein solution, Nguyen et al. [18] proposed a simple combined model, which simultaneously considers pore blockage and cake filtration, to describe the flux declination. Then, in the study [19], the correlation between the steady-state permeate flux and operation parameters was reported. From the steady-state operation equation, optimal design and operation conditions for each particular application could be established.

However, just a few reports on the optimization of membrane processes and cost estimation have been published, or the cost estimation is too general. For example, Wiley et al. [17] optimized the membrane module configurations for brackish water desalination. However, the operation mode is single-pass and only the membrane cost and energy cost were taken into account. Sethi and Wiesner [20] developed the cost model for the removal of natural organic matter, but the study has not conducted the optimization. In membrane technology, the feed and bleed operation mode, which combines the batch and the single-pass configurations, is commonly utilized for continuous full-scale filtration [21, 22]. Therefore, the optimization of a membrane module operated in feed and bleed mode for protein ultrafiltration is considered. The membrane geometry dimensions and operating conditions are design variables in the problem. The system is represented by a set of ordinary differential equations. The objective function is the annual cost, which consists of various types of capital investments and an operating expense. The capital investments are classified into several categories, which are individually correlated to plant scale, particularly the membrane area. The operating expense is the power consumption.

3.1 Process configuration and model calculation

3.1.1 Membrane plant configuration

The configuration of filtration system is continuous feed and bleed, which is shown schematically in **Figure 3**. The notations are summarized in **Table 4**. There are two main pumps in this operation: the feed pump provides the necessary transmembrane pressure, while the recirculation pump maintains the cross-flow rate

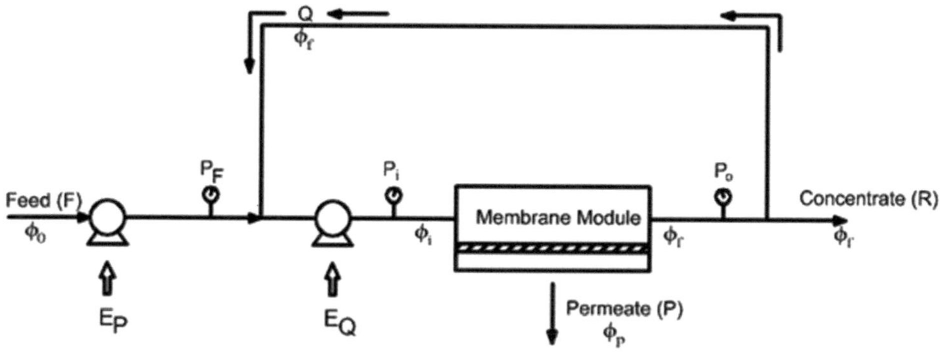

Figure 3.
Schematic configuration of feed-and-bleed mode membrane system.

Notation	Name and units
F	Feed flow rate [m^3/hr]
R	Retentate (concentrate) flow rate [m^3/hr]
Q	Recirculation flow rate [m^3/hr]
Flow	Flow rate in membrane module [m^3/hr]
P	Permeation flow rate [m^3/hr]
P_F	Pressure at outlet of feed pump [kPa]
P_i	Pressure at the inlet of membrane module [kPa]
P_o	Pressure at the inlet of membrane module [kPa]
E_P	Energy consumed by the feed pump [kW]
E_Q	Energy consumed by the recirculation pump [kW]
ϕ_0	Initial concentration of protein solution [m^3/m^3]
ϕ_i	Inlet concentration of protein solution [m^3/m^3]
ϕ_f	Final concentration of protein solution [m^3/m^3]
ϕ_P	Concentration of protein in permeate flux [m^3/m^3]
u	Fluid flow velocity [m/s]
ρ	Fluid density [kg/m^3]
μ	Fluid viscosity [kg/(m·s)]
w, h, L, d_h	Width, height, length, hydraulic diameter of the membrane module

Table 4.
Summary of system configuration notations.

through the modules. The concentrate is continually withdrawn from the system at a flow rate (R).

3.1.2 Modeling of membrane modules

The material balance for total mass and protein give:

$$
\begin{aligned}
F &= R + P \\
F\phi_0 &= R\phi_f + P\phi_p \\
F\phi_0 + Q\phi_f &= (F + Q)\phi_i
\end{aligned}
\tag{15}
$$

The viscosity and density of protein solution correlate to its concentration [23]:

$$
\begin{aligned}
\mu_m &= 8.94 \times 10^{-4} \exp(13.5482\phi_m) \\
\rho_m &= 1000(1 - \phi_m) + 1360\phi_m
\end{aligned}
\tag{16}
$$

in which, μ_m, ϕ_m are the viscosity and concentration of the mixture (solution), respectively. ρ_m is the density of protein solution, 1000 kg/m^3 is the density of water, and 1360 kg/m^3 is the density of dry protein powder.

The permeate flux through the membrane is [19].

$$
\frac{J}{u} = 3.66 \times 10^{-7} \left(\frac{P}{\rho_m u^2}\right)^{0.27} \left(\frac{\rho_m u d_h}{\mu_m}\right)^{0.52} \tag{17}
$$

in which P is the trans-membrane pressure, u is the cross-flow velocity, d_h is the hydraulic diameter of the flow channel.

The equation for permeate flux can be rewritten as:

$$
J = 5.124 \times 10^{-9} \left(\frac{P}{\rho_m u^2}\right)^{0.27} \left(\frac{\rho_m u d_h}{\mu_m}\right)^{0.52} \left(\frac{u}{d_h}\right) \tag{18}
$$

or in terms of shear rate $\dot{\gamma} = \frac{6u}{h} = \frac{12u}{d_h}$, where h is the channel height:

$$
J = 4.27 \times 10^{-10} \left(\frac{P}{\rho_m u^2}\right)^{0.27} \left(\frac{\rho_m u d_h}{\mu_m}\right)^{0.52} \dot{\gamma} \tag{19}
$$

The flow rate/velocity drop and channel length change are calculated from the total mass balance and component balance within the control volume dz across the channel. Protein is assumed to be fully rejected by the membrane:

$$
d(Flow \times \phi_m) = 0 \tag{20}
$$

$$
d(Flow) = -J \times d(wz) = -Jw\,dz \tag{21}
$$

The pressure loss is estimated by the Darcy-Weisbach Equation [24, 25].

$$
dP = -4f\rho_m \frac{dz}{d_h} \frac{u^2}{2} \tag{22}
$$

in which f is the friction factor.

The set of ordinary equations that describes the membrane module system was established as follows [26].

$$\begin{cases} \dfrac{d(Flow)}{d\phi_{\mathrm{m}}} = -\dfrac{Flow}{\phi_{\mathrm{m}}} \\ \dfrac{dz}{d\phi_{\mathrm{m}}} = \dfrac{Flow}{\phi_{\mathrm{m}} wJ} \\ \dfrac{dP}{d\phi_{\mathrm{m}}} = -4f\rho_{\mathrm{m}} \dfrac{1}{d_{\mathrm{h}}} \dfrac{u^2}{2} \dfrac{Flow}{\phi_{\mathrm{m}}} \dfrac{1}{wJ} \\ \mu_{\mathrm{m}} = 8.94 \times 10^{-4} \exp(13.5482\phi_{\mathrm{m}}) \\ \rho_{\mathrm{m}} = 1000(1-\phi_{\mathrm{m}}) + 1360\phi_{\mathrm{m}} \\ J = 5.124 \times 10^{-9} \left(\frac{P}{\rho_{\mathrm{m}} u^2}\right)^{0.27} \left(\frac{\rho_{\mathrm{m}} u d_{\mathrm{h}}}{\mu_{\mathrm{m}}}\right)^{0.52} \left(\dfrac{u}{d_{\mathrm{h}}}\right) \\ u = \dfrac{Flow}{h \times w} \\ f = \dfrac{24}{\mathrm{Re}} \text{ if Re} < 2000 \\ f = \dfrac{0.079}{\mathrm{Re}^{0.25}} \text{ if Re} > 2000 \end{cases} \tag{23}$$

in the range of concentration $[\phi_{\mathrm{i}}, \phi_{\mathrm{f}}]$ and the initial condition

$$\begin{cases} Flow(\phi_{\mathrm{i}}) = F + Q \\ z(\phi_{\mathrm{i}}) = 0 \\ P(\phi_{\mathrm{i}}) = P_{\mathrm{i}} \end{cases} \tag{24}$$

The system of the ordinary equations can be solved numerically by Runge–Kutta fourth-order method [27] to obtain the flow rate, the length, and the pressure. From that, the two important factors determining the total cost, membrane area, and total energy were calculated:

$$A_{\mathrm{membrane}} = w \times L \tag{25}$$

$$\begin{aligned} E_{\mathrm{pumps}} &= E_{\mathrm{P}} + E_{\mathrm{Q}} \\ &= F \times P_i + Q \times \Delta P_{\mathrm{drop}} \end{aligned} \tag{26}$$

In this equation,

E_{P}, E_{Q} are the power supplied by the feed pump and recirculation pump, respectively.

P_{F}, P_{i}, P_{o} are the pressure at the outlet of the feed pump, at the inlet and outlet of the membrane module, respectively.

ΔP_{drop} is the pressure drop in the membrane module.

3.1.3 Cost estimation

3.1.3.1 Operating cost

The operating cost consists of power consumption of the pumps and membrane replacement. The annual energy expense of the pumps is calculated as

$$C_{\text{energy}} = \frac{E_{\text{pumps}}}{\eta} \times 8000 \times \frac{3600}{1000} \times \text{electricity price} \qquad [\$/\text{year}] \qquad (27)$$

E_{pumps} is from Eq. (26). The system is assumed to work 8000 hr./year (24 hr./day and 333 days/year). η is the efficiency of the pumps, which is set to be 0.7. The electricity price is supposed to be 0.08 $/kWh, which is the price for the industrial sector in the United States [28].

The membrane replacement cost is calculated as

$$C_{\text{membrane}} = A_{\text{membrane}} \times c_{\text{membrane}} \times \left(\frac{A}{P}\right) \qquad [\$/\text{year}] \qquad (28)$$

where C_{membrane} [$/m^2/year] is the membrane replacement cost calculated per year.

c_{membrane} [$] is the membrane price per unit area,

$\left(\frac{A}{P}\right)$ is the amortization factor, which presents the time value of money [29] and is calculated as a function of interest rate i and the membrane life.

The membrane price is usually about 200 $/m^2 [9], and membrane life is 12–18 months. Therefore, the membrane replacement cost per year is roughly estimated as 200 $/m^2/year for the interest of i = 8%.

3.1.3.2 Capital cost

It is widely observed that capital costs are correlated to the size in the power-law form [30]:

$$\text{cost} = k(\text{size})^n \qquad (29)$$

In order to achieve higher accuracy, rather than simply predicting the whole capital cost of the membrane plant to capacity, Sethi and Wiesner [20] divided the capital investment into several major categories, which was correlated to the size independently. The major categories include pumps and other manufactured equipment.

a. Pump capital cost

The pumps capital cost can be estimated as (Perry et al. [31]):

$$C^*_{\text{pump}} = I \times f_1 \times f_2 \times C_L \times 81.27 \times (Q \times P)^{0.4} \qquad (30)$$

in which.

I: a cost index ratio for updating the cost to the recent year.

f_1: an adjust factor for pump construction material.

f_2: an adjust factor for suction pressure range.

C_L: a factor used to incorporate labor costs.

Q: flow capacity of the pump [m^3/h].

P: pressure outlet of the pump [kPa].

The cost index, I in Eq. (30), can be referred to as the chemical engineering (CE) index to update the cost. It can be obtained from [32], and the value of 2.4 is used. C_L = 1.4 with the assumption that 40% of the cost is required to install the equipment [33]. The factors f_1 and f_2 can be found in [31]. f_1 = 1.5 when the material is stainless steel. f_2 = 1.0 when the pump pressure is below 10 bar (1 MPa).

The pump size ($Q \times P$ in Eq. (30)) of the two pumps (feed and recirculation pump) is

$$\text{pump size} = (F + Q) \times P_\text{i} \tag{31}$$

b. Capital cost of other equipment

In membrane application, the membrane area is the key parameter, which determines the plant capacity [34]. Thus, the membrane area is chosen as the basic for the estimation of various components in the capital costs.

Non-membrane equipment and facilities, excluding the pumps, were grouped into four main categories: (1) pipes and valves; (2) instruments and controls; (3) tanks and frames; and (4) miscellaneous. The capital cost of each is correlated to the membrane area as follows (Sethi and Wiesner [20])

1. Pipes and valves

$$C^*{}_\text{PV} = 6000(A_\text{membrane})^{0.42} \qquad [\$] \tag{32}$$

2. Instruments and controls

$$C^*{}_\text{IC} = 1500(A_\text{membrane})^{0.66} \qquad [\$] \tag{33}$$

3. Tanks and frames

$$C^*{}_\text{TF} = 3100(A_\text{membrane})^{0.53} \qquad [\$] \tag{34}$$

4. Miscellaneous

$$C^*{}_\text{MI} = 8000(A_\text{membrane})^{0.57} \qquad [\$] \tag{35}$$

c. Annual capital cost

The capital cost can be annualized using the amortization factor as

$$C_\text{capital cost} = C^*{}_\text{capital} \times \left(\frac{A}{P}\right) \qquad [\$/\text{year}] \tag{36}$$

For the plant design year of 20 years and the interest rate 8%, the amortization factor will be about 0.1.

3.2 Formulizations of the problem

3.2.1 Fix parameters and design variables

In the problem, some variables, called input variables, are fixed due to the requirement of the design. In membrane design, these are feed flow F, inlet concentration ϕ_0, and outlet concentration ϕ_f. The protein is assumed to be entirely rejected by the membrane ($\phi_p = 0$).

The design variables were: channel geometry (width × length × height), the inlet pressure (P_i), and recirculation flow rate (Q).

3.2.2 Objective function

The objective function is the sum of capital cost and operating cost, which were annualized:

$$\begin{aligned}\text{minimum} f(\mathbf{x}) &= \text{annual total cost} = \text{annual capital cost} + \text{annual operating cost} \\ &= C_{energy} + C_{membrane} + C_{capital\ (pumps+other)}\end{aligned} \tag{37}$$

3.2.3 Constraints

The pressure at the outlet point should be positive. This constraint is satisfied by assigning a high value to the objective function if the outlet pressure is negative.

The decision variables are frequently limited on a finite range

$$\mathbf{x}_l \leq \mathbf{x} \leq \mathbf{x}_u \tag{38}$$

in which $\mathbf{x}$ is the vector of decision variables $\mathbf{x} = (P_i, Q, w, h)$, subscripts l and u indicate the lower and upper bound.

Parameters	Value
Feed flow rate (m^3/hr)	0.02–200
Inlet pressure (kPa)	200–1000
Recirculation flow rate (m^3/hr)	0–50
Initial solid fraction (m^3/m^3)	0.1
Final solid fraction (m^3/m^3)	0.4
Plant design year (year)	20
Interest rate (%)	8
Energy price ($/kWh)	0.08
Efficiency of pumps (%)	70
Operating temperature (°C)	25
Module height (mm)	0–100
Module width (m)	0–30

Table 5.
System parameters and variables.

The system parameters and variables are summarized in **Table 5**.

3.2.4 Optimization by genetic algorithm

The parameters of GA such as population size, crossover probability, mutation probability values were set to be, 100, 1.0, and 0.30, respectively. The selection was based on roulette wheel with elitism, which means the most fit individual is guaranteed a place in the next generation. The number of generations was assigned to be 500. Because the problem is to minimize the cost, the fitness function was defined as:

$$\text{fitness} = \begin{cases} 0 & \text{if } f(x) > 5 \times 10^5 \\ 5 \times 10^5 - f(x) & \text{otherwise} \end{cases} \tag{39}$$

3.3 Optimum design

For the demonstration of this method, optimum designs of several feed flow rates have been carried out. The lower limit of the membrane width is 0.1 m, the lower limit for the module height is 0.5 mm. The designs are shown in **Table 6**.

Feed [m^3/hr]	ϕ_o [−]	ϕ_f [−]	Pressure [kPa]	Recirculation [m^3/hr]	width [m]	height [mm]	total cost [$/yr]
0.02	0.1	0.4	523	2.8	0.1	5.0	1.29×10^3
0.2	0.1	0.4	1000	4.9	0.1	8.9	4.30×10^3
2	0.1	0.4	1000	0.2	0.1	6.9	1.18×10^4
20	0.1	0.4	987	0.2	1.3	5.0	5.60×10^4
200	0.1	0.4	1000	0.8	11.1	5.0	3.65×10^5

Table 6.
Optimum designs of membrane module.

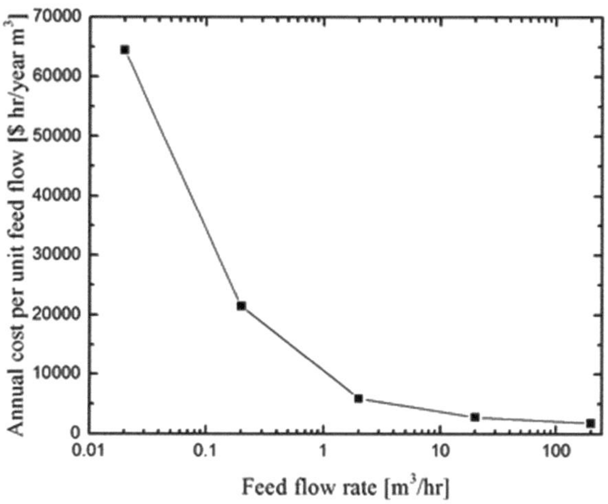

Figure 4.
The behavior of cost per unit flow rate design in optimum condition with plant capacity.

Figure 4 presents the optimum total cost per unit of feed flow. The cost per unit of feed flow decreases with an increase in plant capacity. It reflects the economies of scale.

The results also suggest that the membrane module dimensions and operation condition will change greatly depending on the process requirements, such as the required feed capacity. It is challenging to predict the direction. It might be concluded that the permeate flux also greatly affects the geometric design and operation strategy in membrane separation processes. It is difficult to find a general rule for the design, for each specific system, the correlation between the permeate flux and operating conditions and membrane geometry should be investigated.

4. Modeling and optimization of the BSCF-based single-chamber solid oxide fuel cell by artificial neural network and genetic algorithm

Fuel cells that are highly effective and green technology for converting chemical energy stored in fuel to useable power are currently regarded as one of the most promising approaches for future energy requirements [35]. The solid oxide fuel cell (SOFC) has demonstrated an exceptional integration of advantages, such as high efficiency, fuel flexibility, wide contamination acceptance, and low pollution [36, 37]. Modeling and simulation are valuable tools for determining the impact of various design factors and operating conditions on cell performance, as well as for improving fuel cells [38–40]. Plenty of models have been reported to add to the understanding of fuel cells. Modeling approaches can be categorized into two types: theoretical and empirical one [39, 41, 42]. In the theoretical approach, the spatial dimensions of the models range from simple 0 (0-D) [43, 44] and 1 (1-D) [42, 45–47], to more complicated 2 (2-D) [48–51] and 3 (3-D) [52–54], all with various characteristics and directed at different objectives. The mathematical models, which are based on conservation principles, require a lot of data on parameters and properties of fuel cell, as well as complicated equations and time-consuming calculation.

Empirical or data-driven approach may be more feasible for fuel cell users since the behavior can be quickly and simply deduced without a comprehensive understanding of the internal components, just based on the experimental data [39, 41]. Least squares support vector machine (LS-SVM) [55], Hammerstein model [56, 57] are examples of these approaches. In this approach, artificial neural network (ANN) shows several advantages, including high nonlinearity, rapid computation, a low degree of error in matching experimental data. Using ANNs to model SOFCs appears to be a very promising method.

In this section, an ANN was used to model the performance of the BSCF/GDC-based cathode SOFC. The cell voltage was predicted from cathode sintering temperature, cell operating temperature, and cell current. Several network architectures were examined to find the best structure, and the network was trained using back-propagation methods. The data for training, validation, and testing were taken from our study [58]. The genetic algorithm and the developed ANN were then used to find the best conditions for achieving maximum power.

4.1 Artificial neural network models

Artificial neural networks (ANNs), which were analogous to biological nervous systems, consist of interconnected nodes known as neurons to receive and transfer data [59]. The most basic form, feed-forward architecture, is made up of an input layer, one hidden layers, and an output layer. The input and output layers have the same number of neurons as the number of inputs and outputs in the system to be modeled. Weighted connections connect each neuron to every other neuron in the next layer. In any layer except the input, the weighted sum of data from the previous layer is the input of a neuron. The neuron then activates the data using a function and transfers the response to all neurons in the next layer. The size of the hidden layers is a significant factor that affects the estimation precision because it can make the network become insufficient or overfitting [60]. The number of neurons in hidden layer is generally determined through trials. **Figure 5** illustrates a 3–5-1 feed forward artificial neural network with operating temperature, sintering temperature, and current as inputs.

The activation function employed in this model is the logistic sigmoid$f(x) = 1/(1 + e^{-x})$.

The input data (x_i) are scaled to normalized value (x_{norm}) to enhance the performance of the network:

$$x_{norm} = 0.8 \frac{(x - x_{min})}{x_{max} - x_{min}} + 0.1 \tag{40}$$

in which x_{max} and x_{min} are the bounded interval of the experimental data.

To assess the performance of ANN, the mean squared error (MSE) and coefficient of determination (R^2) are usually used [61].

Various factors affect the performance of fuel cells such as cathode and anode structure, electrolyte material and thickness, cell temperature, inlet and outlet gas compositions. Two important factors, cathode sintered temperature and cell operating temperature, were considered in this model. The sintered temperature is from 1000–1050°C, whereas the operating temperature ranges from 625–700°C. The sintered temperature affects the structure of the obtained cathode as reported in [58]. The explanation of the range for the investigated parameters can be found in [62].

An ANN with one input layer, one hidden layer, and one single output layer was proposed. Current density, sintered temperature of the cathode, and cell operating temperature are the inputs. Back-propagation algorithm [63] was used to train the network. The maximum number of iteration and minimum performance gradient were set to 400 and 10^{-5}, respectively, to stop the training. The proper network structure is determined through a series of trial tests. The data were split into three subsets at random: training, validation, and test, each containing 70, 20, 10% of the

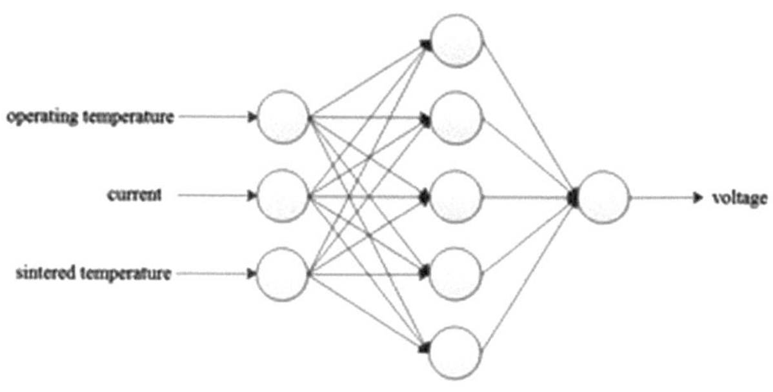

Figure 5.
Artificial neural network (3–5-1) structure.

total samples, respectively. The validation and test sets are necessary for evaluating the validation and power of the networks.

The parameters of the neural network were saved and utilized in the next stage to optimize the power density using genetic algorithms.

4.2 Optimization by genetic algorithm

The objective function is the power density of the fuel cell

$$P = I \times V \tag{41}$$

where P is the power density (mW.cm^{-2}), I is the current density (mA.cm^{-2}) and is the input to the ANN model, and V is the cell voltage (V), which is calculated from the model.

The design variables and their corresponding ranges are summarized as follows:

- sintered temperature of the cathode, [1000–1050] (°C).
- operating temperature of the cell, [625–700] (°C).
- electric current of the cell, [0–1500] (mA.cm^{-2})

The parameters of GA as population size, mutation probability values were set to be 100 and 0.10, respectively. The survival of the individuals was decided by roulette wheel with elitism. The number of generations was 500.

4.3 Optimization results

Figure 6 depicts the fitness values (maximum and mean) of the population versus generation. As indicated in the figure, after about 20 generations, the value of fitness function attained to a maximum value and then remained unchanged. After 100

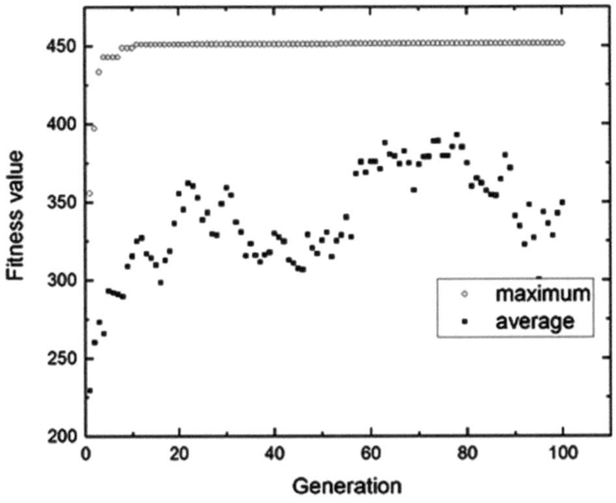

Figure 6.
The fitness values versus generation.

generations, the maximum fuel cell power density of 451.64 mW/cm^2 could be achieved at the sintered temperature of 1005°C, operating temperature of 668°C, and current density of 777 mA/cm^2.

5. Conclusions

The application of genetic algorithm in chemical engineering processes has been illustrated by three case studies. The results suggest that the optimum conditions of complex chemical problems can be easily obtained using genetic algorithm. The successes of genetic algorithm for the challenging problems reported herein, the development of many faster and flexible versions of GA, the improvement of computing ability all suggest the continually increasing impact of metaheuristic methods in chemical engineering systems.

Author details

Thi Anh-Nga Nguyen[1] and Tuan-Anh Nguyen[2]*

1 Faculty of Applied Sciences, Ton Duc Thang University, Ho Chi Minh City, Vietnam

2 Faculty of Chemical Engineering, Ho Chi Minh City University of Technology, VNU-HCM, Vietnam

*Address all correspondence to: anh.nguyen@hcmut.edu.vn

References

[1] Goldberg DE. Genetic Algorithms in Search, Optimization, and Machine Learning. Boston, MA, United States: Addison-Wesley; 1989

[2] Deb K. Introduction to genetic algorithms for engineering optimization. In: Onwubolu GC, Babu BV, editors. New Optimization Techniques in Engineering. Berlin, Heidelberg: Springer Berlin Heidelberg; 2004. pp. 13-51

[3] Nielsen A, Aika K, Christiansen LJ, Dybkjaer I, Hansen JB, Nielsen PEH, et al. Ammonia: Catalysis and Manufacture. Berlin Heidelberg: Springer; 1995

[4] Ammonia AM. Priciples & Industrial Practice. Weinheim, Germany: Wiley; 1999

[5] Babu B, Angira R, Nilekar A. Optimal design of an auto-thermal ammonia synthesis reactor using differential evolution. In: Proceedings of the Eighth World Multi-Conference on Systemics, Cybernetics and Informatics (SCI-2004). Orlando, Florida, USA: International Institute of Informatics and Systemics; 2004

[6] Babu BV, Angira R. Optimal design of an auto-thermal ammonia synthesis reactor. Computers & Chemical Engineering. 2005;**29**(5):1041-1045. DOI: 10.1016/j.compchemeng. 2004.11.010

[7] Carvalho EP, Borges C, Andrade D, Yuan JY, Ravagnani MASS. Modeling and optimization of an ammonia reactor using a penalty-like method. Applied Mathematics and Computation. 2014; **237**:330-339. DOI: 10.1016/j. amc.2014.03.099

[8] Edgar TF, Himmelblau DM, Lasdon LS. Optimization of Chemical Processes. New York, NY, USA: McGraw-Hill; 2001

[9] Ksasy M, Areed F, Saraya S, Khalik MA. Optimal reactor length of an auto-thermal ammonia synthesis reactor. IJECS: International Journal of Electrical and Computer Sciences. 2010;**10**(3):6-11

[10] Murase A, Roberts HL, Converse AO. Optimal thermal Design of an Autothermal Ammonia Synthesis Reactor. Industrial & Engineering Chemistry Process Design and Development. 1970;**9**(4):503-513. DOI: 10.1021/i260036a003

[11] Upreti SR, Deb K. Optimal design of an ammonia synthesis reactor using genetic algorithms. Computers & Chemical Engineering. 1997;**21**(1):87-92. DOI: 10.1016/0098-1354(95)00251-0

[12] Yusup S, Zabiri H, Yusoff N, Yew YC. Modeling and Optimization of Ammonia Reactor Using Shooting Methods. Bucharest, Romania: Proceedings of the 5th WSEAS international conference on Data networks, communications and computers, World Scientific and Engineering Academy and Society (WSEAS); 2006. pp. 258-268

[13] Nguyen TAN, Nguyen TA, Vu TD, Nguyen KT, Dao TKT, Huynh KPH. Optimization of an auto-thermal ammonia synthesis reactor using cyclic coordinate method. IOP Conference Series: Materials Science and Engineering. 2017;**206**:012059. DOI: 10.1088/1757-899x/206/1/012059

[14] Mores PL, Arias AM, Scenna NJ, Caballero JA, Mussati SF, Mussati MC.

Membrane-based processes: Optimization of hydrogen separation by minimization of power, membrane area, and cost. Processes. 2018;**6**(11):221. DOI: 10.3390/pr6110221

[15] Ramírez-Santos ÁA, Bozorg M, Addis B, Piccialli V, Castel C, Favre E. Optimization of multistage membrane gas separation processes. Example of application to CO2 capture from blast furnace gas. Journal of Membrane Science. 2018;**566**:346-366. DOI: 10.1016/j.memsci.2018.08.024

[16] Ohs B, Lohaus J, Wessling M. Optimization of membrane based nitrogen removal from natural gas. Journal of Membrane Science. 2016;**498**: 291-301. DOI: 10.1016/j.memsci. 2015.10.007

[17] Wiley DE, Fell CJD, Fane AG. Optimisation of membrane module design for brackish water desalination. Desalination. 1985;**52**(3):249-265. DOI: 10.1016/0011-9164(85)80036-9

[18] Nguyen T-A, Yoshikawa S, Karasu K, Ookawara S. A simple combination model for filtrate flux in cross-flow ultrafiltration of protein suspension. Journal of Membrane Science. 2012;**403-404**:84-93. DOI: 10.1016/j.memsci.2012.02.026

[19] Nguyen T-A, Yoshikawa S, Ookawara S. Steady state permeate flux estimation in cross-flow ultrafiltration of protein solution. Separation Science and Technology. 2014;**49**(10):1469-1478. DOI: 10.1080/ 01496395.2014.893533

[20] Sethi S, Wiesner MR. Performance and cost modeling of ultrafiltration. Journal of Environmental Engineering. 1995;**121**(12):874-883. DOI: 10.1061/ (ASCE)0733-9372(1995)121:12(874)

[21] Cheryan M. Ultrafiltration and Microfiltration Handbook. Boca Raton, FL, USA: Taylor & Francis; 1998

[22] Zeman LJ, Zydney AL. Microfiltration and Ultrafiltration: Principles and Applications. New York, NY, USA: CRC Press; 1996

[23] Karasu K. A Study on Permeation Phenomena in Cross-Flow Ultrafiltration Producing a Compressible Cake Layer. Tokyo, Japan: Tokyo Institute of Technology; 2010

[24] White FM. Fluid Mechanics. 8th ed. New York, NY, USA: McGraw-Hill Education; 2016

[25] Fox RW, McDonald AT, Mitchell JW. Fox and McDonald's Introduction to Fluid Mechanics. Hoboken, NJ, USA: Wiley; 2020

[26] Nguyen T-A, Yoshikawa S. Modeling and economic optimization of the membrane module for ultrafiltration of protein solution using a genetic algorithm. Processes. 2020;**8**(1):4. DOI: 10.3390/pr8010004

[27] Dorfman KD, Daoutidis P. Numerical Methods with Chemical Engineering Applications. Cambridge, United Kingdom: Cambridge University Press; 2017

[28] U.S. Energy Information Administration. Average Retail Price of Electricity to Ultimate Customers Total End-Use Sector. U.S., USA: Department of Energy; 2019

[29] Park CS. Fundamentals of Engineering Economics. London, United Kingdom: Pearson Education; 2013

[30] Turton R, Bailie RC, Whiting WB, Shaeiwitz JA. Analysis, Synthesis and Design of Chemical Processes. London,

United Kingdom: Pearson Education; 2008

[31] Perry RH, Perry RH, Chilton CH, Perry JH. Chemical Engineers' Handbook. 5th ed. New York, NY, USA: McGraw-Hill; 1973

[32] Green DW, Perry RH. Perry's Chemical Engineers' Handbook. Eighth ed. New York, NY, USA: McGraw-Hill Education; 2007

[33] Holland FA, Wilkinson JK. Process economics. In: Perry RH, Green DW, editors. Perry's Chemical Engineers' Handbook. 7th ed. New York, NY, USA: McGraw-Hill; 1997

[34] Mir L, Michaels SL, Goel V, Kaiser R. Crossflow Microfiltration: Applications, Design, and Cost. In: Ho WSW, Sirkar KK, editors. Membrane handbook: Newyork, NY, USA: Springer; 1992. pp. 571-594

[35] Wachsman ED, Marlowe CA, Lee KT. Role of solid oxide fuel cells in a balanced energy strategy. Energy & Environmental Science. 2012;**5**(2): 5498-5509. DOI: 10.1039/C1EE02445K

[36] Ghezel-Ayagh H, Borglum BP. (invited) review of Progress in solid oxide fuel cell at FuelCell energy. ECS Transactions. 2017;**80**(9):47-56. DOI: 10.1149/08009.0047ecst

[37] Minh N, Mizusaki J, Singhal SC. Advances in solid oxide fuel cells: Review of Progress through three decades of the international symposia on solid oxide fuel cells. ECS Transactions. 2017;**78**(1):63-73. DOI: 10.1149/07801.0063ecst

[38] Kakaç S, Pramuanjaroenkij A, Zhou XY. A review of numerical modeling of solid oxide fuel cells. International Journal of Hydrogen Energy. 2007;**32**(7):761-786. DOI: 10.1016/j.ijhydene.2006.11.028

[39] Hajimolana SA, Hussain MA, Daud WMAW, Soroush M, Shamiri A. Mathematical modeling of solid oxide fuel cells: A review. Renewable and Sustainable Energy Reviews. 2011;**15**(4): 1893-1917. DOI: 10.1016/j.rser.2010.12.011

[40] Ma L, Ingham DB, Pourkashanian M, Carcadea E. Review of the computational fluid dynamics modeling of fuel cells. Journal of Fuel Cell Science and Technology. 2005;**2**(4): 246-257. DOI: 10.1115/1.2039958

[41] Wang K, Hissel D, Péra MC, Steiner N, Marra D, Sorrentino M, et al. A review on solid oxide fuel cell models. International Journal of Hydrogen Energy. 2011;**36**(12):7212-7228. DOI: 10.1016/j.ijhydene.2011.03.051

[42] Karcz M. From 0D to 1D modeling of tubular solid oxide fuel cell. Energy Conversion and Management. 2009; **50**(9):2307-2315. DOI: 10.1016/j.enconman.2009.05.007

[43] Costamagna P, Magistri L, Massardo AF. Design and part-load performance of a hybrid system based on a solid oxide fuel cell reactor and a micro gas turbine. Journal of Power Sources. 2001;**96**(2):352-368. DOI: 10.1016/S0378-7753(00)00668-6

[44] Zabihian F, Fung AS. Macro-level modeling of solid oxide fuel cells, approaches, and assumptions revisited. Journal of Renewable and Sustainable Energy. 2017;**9**(5):054301. DOI: 10.1063/1.5006909

[45] Ota T, Koyama M, Wen C-j, Yamada K, Takahashi H. Object-based modeling of SOFC system: Dynamic behavior of micro-tube SOFC. Journal of

Power Sources. 2003;**118**(1):430-439. DOI: 10.1016/S0378-7753(03)00109-5

[46] Li P-W, Suzuki K. Numerical modeling and performance study of a tubular SOFC. Journal of The Electrochemical Society. 2004;**151**(4): A548. DOI: 10.1149/1.1647569

[47] Bove R, Lunghi P, M. Sammes N. SOFC mathematic model for systems simulations—Part 2: Definition of an analytical model. International Journal of Hydrogen Energy. 2005;**30**(2): 189-200. DOI: 10.1016/j.ijhydene. 2004.04.018

[48] Ma R, Gao F, Breaz E, Huangfu Y, Briois P. Multidimensional reversible solid oxide fuel cell modeling for embedded applications. IEEE Transactions on Energy Conversion. 2018;**33**(2):692-701. DOI: 10.1109/ TEC.2017.2762962

[49] Ni M. 2D thermal-fluid modeling and parametric analysis of a planar solid oxide fuel cell. Energy Conversion and Management. 2010; **51**(4):714-721. DOI: 10.1016/j. enconman.2009.10.028

[50] Geisler H, Dierickx S, Weber A, Ivers-Tiffee E. A 2D stationary FEM model for hydrocarbon Fuelled SOFC stack layers. ECS Transactions. 2015; **68**(1):2151-2158. DOI: 10.1149/ 06801.2151ecst

[51] Luo XJ, Fong KF. Development of 2D dynamic model for hydrogen-fed and methane-fed solid oxide fuel cells. Journal of Power Sources. 2016;**328**: 91-104. DOI: 10.1016/j.jpowsour. 2016.08.005

[52] Nikooyeh K, Jeje AA, Hill JM. 3D modeling of anode-supported planar SOFC with internal reforming of methane. Journal of Power Sources. 2007;**171**(2):601-609. DOI: 10.1016/j. jpowsour.2007.07.003

[53] Andersson M, Paradis H, Yuan J, Sundén B. Three dimensional modeling of an solid oxide fuel cell coupling charge transfer phenomena with transport processes and heat generation. Electrochimica Acta. 2013;**109**:881-893. DOI: 10.1016/j.electacta.2013.08.018

[54] Yang C, Yang G, Yue D, Yuan J, Sunden B. Computational fluid dynamics model development on transport phenomena coupling with reactions in intermediate temperature solid oxide fuel cells. Journal of Renewable and Sustainable Energy. 2013;**5**(2):021420. DOI: 10.1063/1.4798789

[55] Huo H-B, Zhu X-J, Cao G-Y. Nonlinear modeling of a SOFC stack based on a least squares support vector machine. Journal of Power Sources. 2006;**162**(2):1220-1225. DOI: 10.1016/j. jpowsour.2006.07.031

[56] Huo H-B, Zhong Z-D, Zhu X-J, Tu H-Y. Nonlinear dynamic modeling for a SOFC stack by using a Hammerstein model. Journal of Power Sources. 2008;**175**(1):441-446. DOI: 10.1016/j.jpowsour.2007.09.059

[57] Jurado F. A method for the identification of solid oxide fuel cells using a Hammerstein model. Journal of Power Sources. 2006;**154**(1):145-152. DOI: 10.1016/j.jpowsour.2005.04.005

[58] Le M-V, Tsai D-S, Nguyen T-A. BSCF/GDC as a refined cathode to the single-chamber solid oxide fuel cell based on a LAMOX electrolyte. Ceramics International. 2018;**44**(2): 1726-1730. DOI: 10.1016/j.ceramint. 2017.10.103

[59] Baughman DR, Liu YA. Neural Networks in Bioprocessing and Chemical

Engineering. Amsterdam, Netherlands: Elsevier Science; 1995

[60] Sheela KG, Deepa SN. Review on methods to fix number of hidden neurons in neural networks. Mathematical Problems in Engineering. 2013;**2013**:425740. DOI: 10.1155/2013/425740

[61] Himmelblau DM. Accounts of experiences in the application of artificial neural networks in chemical engineering. Industrial & Engineering Chemistry Research. 2008;**47**(16): 5782-5796. DOI: 10.1021/ie800076s

[62] Le M-V, Nguyen T-A, Nguyen TA-N. Modeling and optimization of the BSCF-based single-chamber solid oxide fuel cell by artificial neural network and genetic algorithm. Journal of Chemistry. 2019;**2019**:7828019. DOI: 10.1155/2019/7828019

[63] Wythoff BJ. Backpropagation neural networks: A tutorial. Chemometrics and Intelligent Laboratory Systems. 1993; **18**(2):115-155. DOI: 10.1016/0169-7439(93)80052-J

Chapter 6

Using Genetic Algorithm to Optimize Controllers of Thermal Load System in Thermal Power Plant

PhamThi Ly and Bui Quoc Khanh

Abstract

This chapter presents the sequence of implementing the genetic algorithms using the programming language in the Mfile application of MATLAB Simulink software to optimize the two controller parameters of the coordinated control system structure of the thermal load system in coal-fired thermal power plants: electric power controller and steam pressure controller. Optimal standards are determined to be fast-tracking and fuel-saving. Operational data at a thermal power plant in Vietnam have been used to simulate the operation of a thermal load control system with a coordinated control structure in a thermal power plant to test controller parameters found from the genetic solution. To clarify the superiority of the genetic algorithm method in control of the thermal load system of a thermal power plant, the authors give an evaluation of the original control system compared to the control system using the parameters found from the genetic algorithm method. The results show that the thermal load control system in the thermal power plant using controller parameters found from the genetic algorithm method is much more optimal in terms of fuel consumption and the ability to follow the set amount.

Keywords: genetic algorithm, coordinated control system, thermal power plant, thermal load control system, optimization

1. Introduction

Genetic algorithms are general algorithms that can successfully solve difficult problems in many fields, which cannot be solved by other methods [1–6]. The application scope of genetic algorithms is confronted with problems that are impossible or ineffective to solve.

In recent years, the applications of genetic algorithms have increased greatly in many fields such as engineering (engine design, aircraft design, etc.), optimization, robot operation, product classification system, machine learning system, pattern recognition, neural network training, fuzzy system tuning, planning, adaptive control, game programming, transportation problems, tourism problems, etc. [1–3].

The control system's design involves many issues, such as, system stability, transient and steady-state quality, etc. Each problem depends a lot on the structure and parameters of the control system. However, this dependence may not be expressed mathematically. In addition, when designing, it is necessary to ensure that conflicts among the quality criteria are adequately resolved. The lack of systematic methods for selecting values for many controller parameters is a major obstacle to satisfying control requirements. To solve these problems using genetic algorithms, we encode the structure and parameters of the controller into a chromosomal sequence and define a fitness function as a function of the quality requirements. Accordingly, we can convert the design problem into the problem of minimizing an objective function according to the controller parameters. Since genetic algorithms only use the fitness function in the optimization process, they can perform this search. The innovative combination of existing control methods with genetic algorithms can create a powerful tool to solve real control problems.

The field of control of coal-fired power plants is complicated and contains many difficult processes to solve, especially the thermal load control system of coal-fired power plants [7–9]. Currently, many authors have used genetic algorithms in their research to solve problems in production and control activities at coal-fired power plants. The studies [10–15] all have a thermal load system model described by a first-order differential equation or a system of nonlinear equations. These control models have ignored the physical effects of the boiler base loops in the thermal load system in the thermal power plant: the fuel control loop, the gas and air control loop, the water supply control loop, superheat steam, and spray control loop. The use of genetic algorithms in these studies is mainly using the GA toolbox in MATLAB Simulink with different goals: [10–12] interested in reducing fuel consumption in the production of each MW of electricity, fast-tracking the setpoint; [16] interested in minimizing the influence of turbine valve adjustment and combustion speed on the power generator; [13, 17] investigated all the objectives which are related to each parameter in the operation of thermal power plants: coal flow, smoke flow, ID fan, FD fan, heat reduction spray flow of the process of superheating and spray, etc. while [14] gives the target of the fastest action time for the control parameters of the fuel valve, steam valve, and water supply valve.

In principle, the parameters of a PID controller in a control system can be easily designed using classical methods such as root locus or bode plots, etc. However, in the industry, most parameters of the PID controller are experimentally adjusted by the errors test method because there are no mathematical models of the object. The adjustment process in many cases is very difficult and time-consuming. A proposed solution to solve the above-mentioned difficulties is to use a genetic algorithm to automatically adjust the parameters of the PID controller so that the control systems reach the minimum value of criterion.

In this chapter, a genetic algorithm is used to optimize the parameters of two main PID controllers in the thermal load control system with a new coordinated structure which is proposed by the authors (G_{CN} power controller and G_{CP} steam pressure controller). This new coordinated control structure built by the authors is based on a method that combines the description of dynamic processes with data collection and operating parameters of a real thermal power plant. This is a complicated model, including many linear and nonlinear processes with mutual and interleaving influence. The goal is used to find solutions for optimal parameters of power controller and pressure controller using genetic algorithms is to optimize the operation: fast-tracking the setpoint and saving the fuel (kg of coal/kWh). Because the model is so large and complicated that no equation can fully describe its characteristics, it is impossible to use the GA toolbox as the other authors do to apply genetic algorithms in finding the optimal parameters for the controllers of the thermal load system. The present authors build a program on MATLAB Simulink's Mfile with all the properties and characteristics of a genetic algorithm. This program will find out the optimal parameters of the controllers in the thermal load system by affecting the model built on Simulink to find the optimal parameters for the controllers.

Achieve expected results:

- A software program will be written for a genetic algorithm is applied to find optimal parameters for the controllers of a model with a complicated structure.
- New optimal parameters are better for the controllers of the thermal load system in thermal power plants.
- The consumption of fuel is reduced during the production of electricity (at least 1 g coal per kW power).

2. New coordinated control structure

2.1 New coordinated control structure of thermal load control system in thermal power plant

The thermal load control system according to the new coordinated structure proposed by this chapter's authors in **Figure 1**:

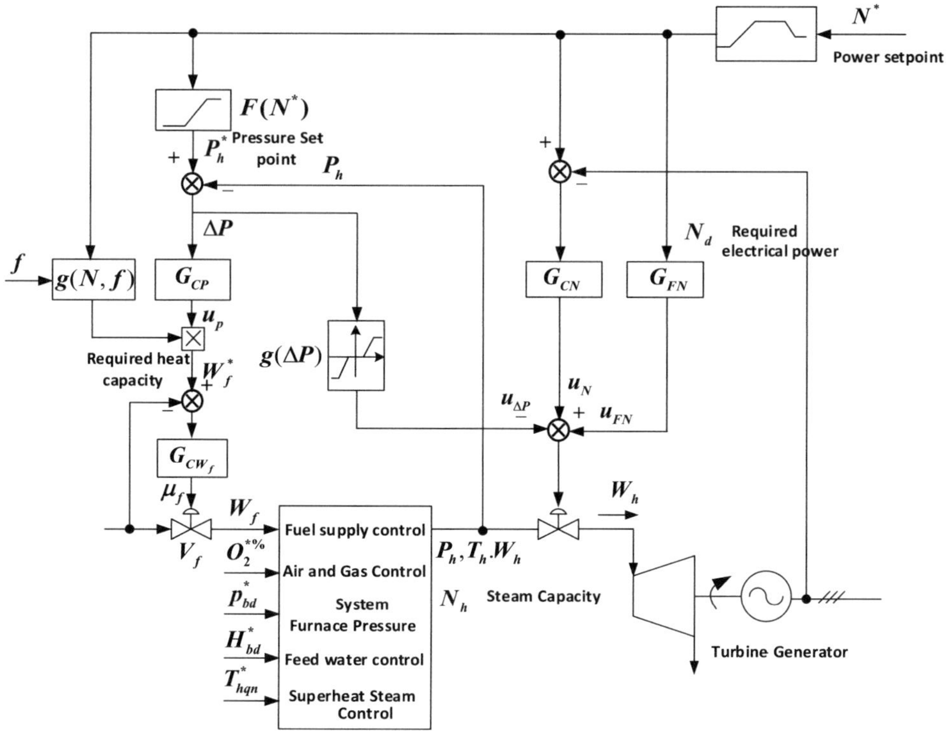

Figure 1.
The new coordinated control structure.

On the new coordinated control structure, the setpoint of power Ne* (MW) is used as the setpoint for both the steam power control system (boiler) and the generating capacity control system (turbine-generator). In the steam power control system, the variables to be controlled are steam temperature and steam pressure. To meet the control requirements of this control system, it is necessary to adjust the basic control loops so that these control loops work well, meeting the operating requirements: fuel control loop, gas and air control loop (control of residual oxygen concentration and control of negative pressure in the combustion), water supply control loop, superheated steam and spray control loop. A pressure regulating limiter is designed to be placed at the outlet of the G_{CP} pressure controller. This setting is to avoid the amount of fuel being too large or zero to cause fluctuations in the amount of fuel set.

On the power generation control system with G_{CN} power controller, the variable that needs to be controlled is the electrical power Ne (MW), and the actuation variable is the turbine regulating valve which is to bring the standard steam flow into the turbine to rotate the machine generate electricity. At the control system side of the turbine-generator cluster, it is designed to feedforward the power setpoint G_{FN} in the turbine control part to increase the ability to fast track the power setpoint. This structure has a signal that compensates for pressure acting on the turbine valve opening signal to eliminate the channeling between the steam capacity and electrical power control systems.

Because fuel often changes the chemical composition and calorific value of coal varies with the type of coal. The determination of variation is usually done by the chemistry lab at the plants that experiment and test these fuels and then set the operating requirements for that coal. In this new coordinated control mechanism, we proposed to add fuel characteristics g (N, f) to actively change the fuel setpoint according to the change of the power setpoint corresponding to each coal with different calorific values, reducing adjustment time and contributing to fuel economy. The characteristic G (N, f) function tested when changing the calorific value of coal affects electrical power and vapor pressure is shown in **Figure 2**.

Due to the large inertia of the boiler, the steam pressure control system responds more slowly to the turbine-generator output steam power requirements to meet the

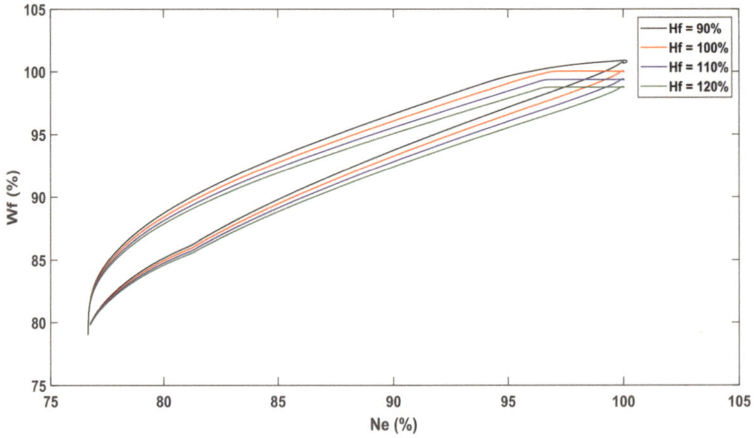

Figure 2.
The G (N, f) characteristic of the new coordination control structure.

capacity electrical requirements. Thus, an inter-channel signal is required between the two vapor pressure control systems and the power control systems. Therefore, in the structure of **Figure 1**, we have introduced a static nonlinear function g(ΔP) [18] to compensate for the interleaving between the two steam pressure control systems G_{CP} and the power control system G_{CN} to ensure that the pressure is slightly more stable and the transmitter power responses faster. This static nonlinear compensation function g(ΔP) is developed from the structure of Flynn [18].

2.2 Simulation of the operation of a thermal load control system with a coordinated control structure

MATLAB Simulink software is used to simulate the control operation of the coordinated control structure in **Figure 1**.

2.2.1 Simulation parameters

Simulation parameters are taken at the stable working time (load from 230 to 300 MW) of a unit with a capacity of 300 MW of the Hai Phong thermal power plant [7, 8].

The volume of the furnace: 8485 m^3.

- Temperature of smoke exiting from the furnace: 1047°C.
- Smoke fan speed: 620 RPM.
- Total flow of air supplied to the boiler: 295.486 kg/s.
- Specific heat capacity of wind: 1005 J/kg.K.
- Coal flow rate: 28.3 kg/s.
- Specific heat of fuel: 1260 J/kg.K.
- Smoke/wind ratio: 1.1.
- Ratio of fuel inlet and slag out: 0.2.
- Coal and wind temperature level 1: 228°C.
- Wind temperature level 2: 341°C.
- Specific heat capacity of water: 4200 J/kg.K.
- Water supply pump speed: 4842 RPM.
- Saturated steam specific heat: 1840 J/kg.K.
- Main steam flow: 191 kg/s.
- Saturated steam pressure: 14.2 MPa.

- Saturated steam temperature: 340°C.
- Main steam temperature: 541°C.
- Water supply flow: 176 kg/s.
- Feedwater temperature: 280°C.
- Water flow to reduce temperature: 18 kg/s.
- Water spray valve opening to reduce heat: 29.42%.
- Water temperature reduced: 25°C.
- Valve opening: 75%.

2.2.2 Simulation model

After building a simulation model of the control loops in the control systems: the boiler control system and the turbine-generator combination control system. The new coordinated control structure of the thermal load control system is shown in **Figure 1**, using experimental data experience and operation collected from Hai Phong thermal power plant [7] to build a simulation model on MATLAB Simulink in **Figure 3**.

Figure 4 shows the coordinated control structure in the thermal load control system:

2.2.3 Simulation results of the control system without changing the coal calorific value

Simulation scenario: The system is operating stably at the balanced point at the generating capacity of 230 MW. There is a request to increase capacity from 230 to 300 MW at 1000 s. At 9000 s, there is a request to reduce the generating capacity from 300 to 230 MW. Load increase and decrease rate is 3 MW/min, steam pressure is set according to sliding pressure characteristic, at 1000 s start to increase from 14.2 to 16.7 MPa (68 ÷ 80%, respectively) and at the time of 9000 s start to decrease from 16.7 to 14.2 MPa.

Figure 5 illustrates the responses of power (Ne), steam pressure (Ph), steam flow (W_h), coal fuel flow (W_f), boiler water level (H), superheated steam temperature (T_h),

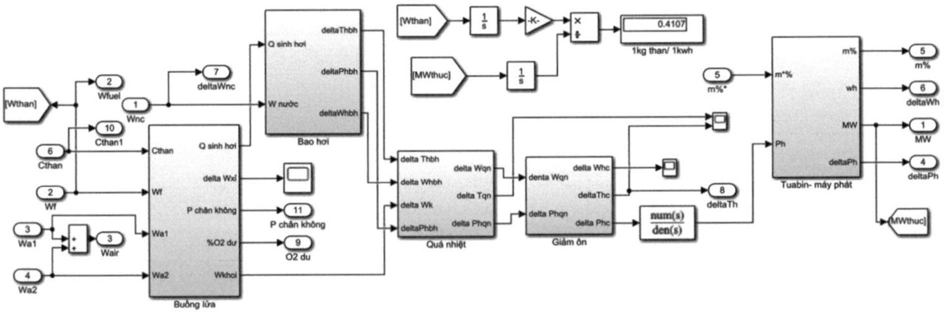

Figure 3.
The model of a thermal load control system in MATLAB Simulink.

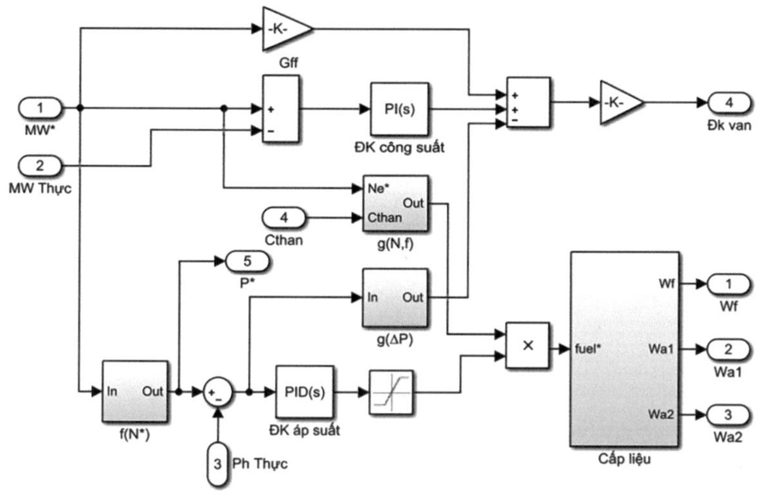

Figure 4.
Coordinated control structure of thermal load system in MATLAB Simulink.

combustion chamber negative pressure (P_{bd}), residual $\%O_2$ concentration (residual O_2) when no change is simulated in the calorific value of coal.

The criteria to evaluate the power and pressure response in **Figure 5** are transient time (T_{qd}), power error, maximum pressure difference (e_{max} (%)) during load increase, decrease time, and static error. We have the following quality rating as shown in **Table 1**.

Thus, it shows that: the power and pressure responses have a fast transition time during both load increase and decrease. The error in the process of increasing and

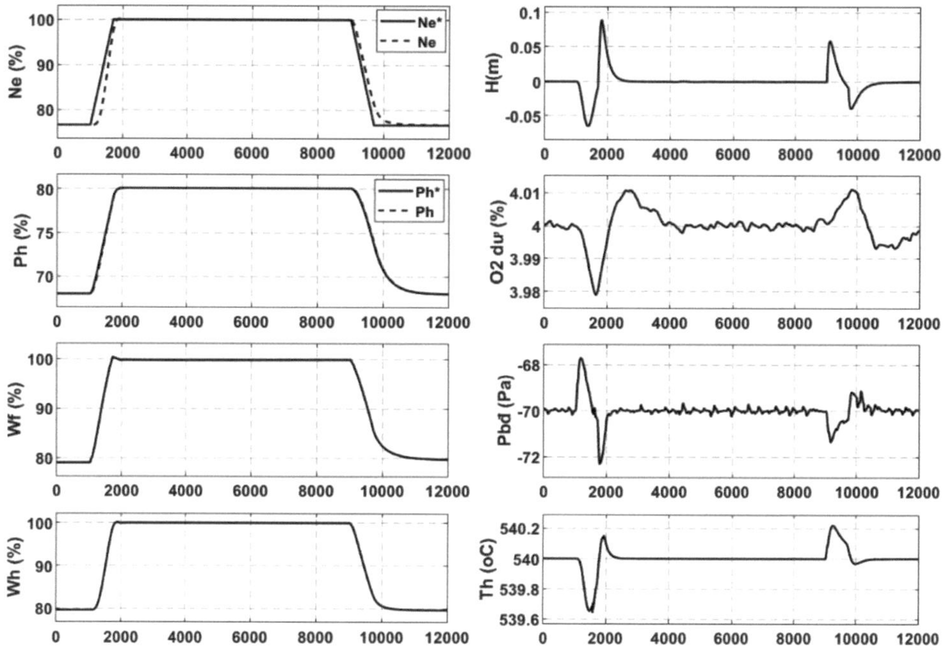

Figure 5.
Simulated application response mode control combinations.

Control loops	T_{qd} (s)		e_{max} (%)	
	Increase load	Reduce load	Increase load	Reduce load
Power response	822	1155	8.06	4.94
Pressure response	812	2250	0.392	0.18

Table 1.
Power and pressure response control quality assessment.

decreasing the load is small. There is no over-adjustment or oscillation. The real power line (Ne) follows the setpoint power line (Ne*), the real pressure line (Ph) follows the setpoint pressure line (Ph*).

The characteristics of fuel flow and steam flow have been met according to control requirements. The four basic control loops in the boiler were working true. The oscillations of the responses around the preset point are within the allowable range. The level of water in the boiler fluctuates in the range of $-65 \div 88$ (mm), the concentration of residual %O_2 fluctuates in the range of $-0.02 \div 0.01$ (%), the negative pressure of the combustion chamber fluctuates in the range of $-67.7 \div 72.3$ (Pa), superheated steam temperature ranges $-0.34 \div 0.22$ (°C).

In summary, the four control loops in the boiler control are working true. The coordinated control structure has also shown good quality when ensuring fast response of pressure control and power control to load requirements, eliminating the inter-channel effect between boiler and turbine-generator.

2.2.4 Simulation results of the control system when the coal calorific value changed

Figure 6 shows the responses of the calorific value of coal (H_{than}), fuel flow control loop (W_f), power error (ErrNe), pressure error (ErrPh), residual %O_2 concentration (residual O_2) control loop, and the combustion chamber negative pressure control loop (Pbd).

Coal calorific value is changed when the system is stable at 100% load capacity. At the time 4000 s start to change the calorific value as shown in **Figure 6**. There are two stages of heating value increase: the heating phase increases from 100% up to 109.5% compared to the average calorific value, and the period of reducing the calorific value to 95.2% compared with the average calorific value.

It can be seen that the fuel response changes immediately as soon as the calorific value of the coal changes. As the calorific value of coal increases, the fuel flow is reduced by 1.15% of the maximum required fuel flow. As the calorific value of coal decreases, the fuel flow increases to 0.66% of the maximum fuel flow.

Power and pressure errors are small when the coal calorific value changes. This error value is so small that it can be ignored. The response of the control loops the residual %O_2 concentration, and the combustion chamber negative pressure also fluctuates slightly around the preset value. It shows that these control loops operate stably, not affected by the interference of temperature change.

Summarizing this section, the combined control structure gives good control quality and works well when there is a change in coal calorific value, which affects the new coordinated control system proposed by the authors. It has eliminated the inter-channel effect between the boiler and the turbine generator.

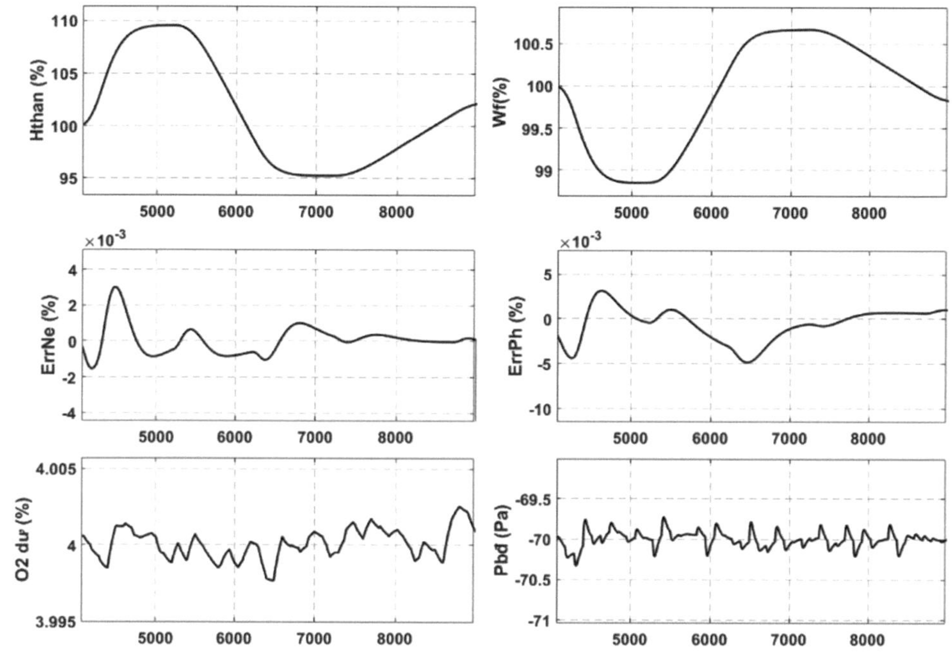

Figure 6.
Simulation response when the coal calorific value changed.

3. Using genetic algorithm to optimize parameters of two controllers of the new coordinated control structure

Although the pressure controller G_{CP} and the power controller G_{CN} have met the control requirements for the thermal load control system of the thermal power plant. However, those controllers only use the IMC-PID method and PID self-tuner tool on MATLAB Simulink. The parameters of the pressure controller and power controller in the new coordinated control structure of the thermal load control system are not the most optimal. Therefore, to improve production efficiency and optimize operation (fast-tracking the setpoints and saving fuel), the chapter's authors continue to research and design a software program using genetic algorithms to find out the optimized parameters of PI and PID controllers of steam pressure control system G_{CP} and power control system G_{CN}.

For the objective of power tracking control, absolute value integral standard is applied with minimal power error as shown in Eq. (1):

$$J_N = \frac{1}{t - t_0} \int_{t_0}^{t} \left| e_N^{\%} \right| dt \rightarrow \min \tag{1}$$

For the target of reducing fuel costs for electricity production (kg coal/kWh), it is calculated as in Eq. (2):

$$J_f = \frac{\int_{t_0}^{t} W_f(t)dt}{\int_{t_0}^{t} N_e(t)dt} \to \min \quad (2)$$

The objective function of refining G_{CN} and G_{CP} controllers using genetic algorithms in this contents are:

$$J = J_N + J_f \quad (3)$$

The purpose of the applied GA algorithm is to find the optimal values {Kp1_out, KI1_out, Kp2_out, KI2_out, Kd2_out} of the controller G_{CN} and G_{CP} of the new coordinate system, where the functions J_N and J_f reaches the minimum value. In other words, the objective function of the GA algorithm is min (J).

To limit the search space of the genetic algorithm, it is assumed that the optimal values {Kp1_out, KI1_out, Kp2_out, KI2_out, Kd2_out} are located around the value {Kp1, Ki1, Kp2, Ki2, Kd2} of two controllers G_{CN} and G_{CP}. We know that these parameters {Kp1, Ki1, Kp2, Ki2, Kd2} are obtained from using the PID Turner tool of MATLAB Simulink and the IMC_PID method. The search limits for those five parameters of the PID controller are as shown in Eq. (4) as follows:

$$\begin{aligned} &\alpha K_{p1} \le K_{p1-out} \le \beta K_{p1} \\ &\alpha K_{i1} \le K_{i1-out} \le \beta K_{i1} \\ &\alpha K_{p2} \le K_{p2-out} \le \beta K_{p2} \\ &\alpha K_{i2} \le K_{i2-out} \le \beta K_{i2} \\ &\alpha K_{d2} \le K_{d2-out} \le \beta K_{d2} \end{aligned} \quad (4)$$

In which, the coefficients α, β are chosen so that the search space is large enough to accommodate the desired optimal value. Simulation results on the thermal load system with the new coordinated control structure, ά = 0 and β = 300 are satisfied.

The genetic algorithm is used as a tool to solve the optimization problem, to obtain the values {Kp1_out, KI1_out, Kp2_out, KI2_out, Kd2_out} satisfying the objective functions on Eq. (3) with space search is limited by Eq. (4). Algorithm flowchart of genetic algorithm to determine PID parameters of two G_{CN} power controllers and G_{CP} pressure controllers in the new coordinated control system in **Figure 7**:

The steps to search for optimal parameters for two controllers G_{CP} and G_{CN} according to the algorithm flowchart of the genetic algorithm method in **Figure 7** are as follows.

3.1 Startup

Individuals in the origin natural population Kp1, Ki1, Kp2, Ki2, Kd2 of the two controllers G_{CP} and G_{CN} are randomly startup so that at the time of initialization they can exist for a long time in the environment. The initial parameters for the startup process are as follows:

- Number of individuals in the initial population: 25

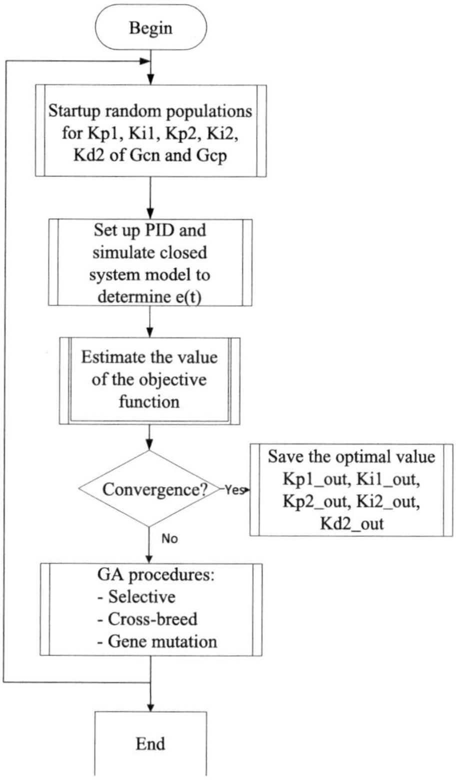

Figure 7.
Process flow chart of genetic algorithm determining PID parameters of two power and pressure controllers in a new coordinated control system.

- Each individual has five chromosomes. Each of these individuals is equivalent to a power controller parameter (Kp1, Ki1) and pressure controller parameters (Kp2, Ki2, Kd2) with a limit of parameters Kp1, Ki1, Kp2, Ki2 from 0 to 10 and Kd2 parameter from 0 to 300. The subroutine of the startup process is as follows:

```
function par = Init(N,d,range)
% input variables:
% N: population size
% d: Number of parameters of the function to find the extreme
% range: two-dimensional array [2xd] stores the value domain of the parameters
% output variables:
% par: two-dimensional array [Nxd]store randomly initialized population
for pop_index = 1:N
  for par_index = 1:d,
        par(pop_index,par_index) = ...
             (rand-0.5)*(range(2,par_index)-
range(1,par_index))+
             0.5*(range(2,par_index) + range(1,par_index));
  end
end
```

3.2 Evaluate

To evaluate individuals, a fitness function must be defined. A fitness function is usually a function that needs to find an extreme or an equivalent transformation of the function. The fitness function is calculated from the objective function (Eq. (5)) as follows:

$$J = (\mathrm{Ne}^* - \mathrm{Ne})^2 + (\mathrm{Ph}^* - \mathrm{Ph})^2 + \left(\frac{w_f}{u_T}\right)^2 \to \min \tag{5}$$

The fitness function is the inverse of the objective function (J):

$$fitness = \frac{1}{J + C} \tag{6}$$

Where C is added to ensure that the fitness function is always positive.

To find the solution of the objective function, we take the parameters Kp1, Ki1, Kp2, Ki2, Kd2 of the two controllers G_{CP} and G_{CN} through the initial start-up process, and put them into the simulation on Simulink, which has built this condition control of the new coordinated control structure. If the PID parameter satisfies the stable system, then GA calculates J. Otherwise, if the system is unstable, stop the simulation and find another PID parameter. Because we use the order of fitness of chromosomes ascending (GA is looking for the max value), we must inverse the objective function to find the max value (if J finds the min). Then the best instance (with the smallest J) is stored in the bestchrom memory cell. The program implemented for the evaluation process is as follows:

```
% Calculate the fitness of the initial population
for pop_index = 1:N,
    Kp1 = par(pop_index,1);
    Ki1 = par(pop_index,2);
    Kp2 = par(pop_index,3);
    Ki2 = par(pop_index,4);
    Kd2 = par(pop_index,5);
    sim('newcoordinated.slx');% Simulation of the
                              % coordinated control
                              % system in figures 3 and 4
 if length(e1) > 79500
        Kp1;
        Ki1;
        Kp2;
        Ki2;
        Kd2;
    J = (e1'*e1) + (e2'*e2) + (e3'*e3);
    fitness(pop_index) = 1/(J + eps);
 else
        J = 10^100;
        fitness(pop_index) = 1/(J + eps); % GA to find the
                                          % maximum of the
```

```
                                        % adaptive function
  end
 end;
 [bestfit,bestchrom] = max(fitness)
```

Thus, after the startup and evaluation of the genetic algorithm, 25 individuals were initially randomly selected to exist in the habitat. These individuals are taken to the next evolutionary process.

3.3 Encode

To apply the genetic algorithm to the optimization problem, it is necessary to encode the solution of the problem into a chromosome sequence. In this problem, we need to find five parameters Kp1, Ki1, Kp2, Ki2, and Kd2. Each solution consisting of five parameters is called an individual containing a sequence of chromosomes. Each parameter is a gene segment on the chromosome. There are three ways commonly used to encode: binary, decimal, and real number encoding. In this problem, the encoding is a decimal encoding.

The genome used to encode the solution in this decimal encoding method consists of 10 symbols {0,1,2,3,4,5,6,7,8,9}. This decimal encoding has the advantage that the NST string length is significantly shortened compared to the binary encoding, thus making the algorithm run faster. It is possible to directly apply the familiar genetic operations given to binary encoding. The implementation program for the decimal encoding process is as follows:

```
% Endcode_Decimal.m: Decimal encoding subroutine
% Programmer: Pham Thi Lý, UTC, HaNoi, VietNam
function pop = Encode_Decimal(par,sig,dec)
% input variables:
% par: two-dimensional array [Nxd]stores the parameters to be encoded
% N: Population size, d is the number of parameters
% sig: row vector [1xd] stores the number of significant digits equivalent to
each parameter
% dec: row vector [1xd] saves decimal point position
% output variables:
% pop: two-dimensional array [NxL] save population after decimal encoding
% each row of pop is an instance (L is the length of the chromosomal string)
function pop = Encode_Decimal(par,sig,dec)
if (nargin <3)
   error(['missing arguments. syntax: pop = Encode_Decimal(par,sig,dec)']);
end;
if size(sig) ~ =size(dec),
   error(['The sig and dec arguments don't match']);
end;
[N,d] = size(par); % Determine population size
for pop_index = 1:N, % Scan from the beginning to the end of the
```

```
                    % population
gene_index = 1;
for par_index = 1:d, % scan parameters
% The first gene encodes the sign of the parameter
if par(pop_index,par_index) < 0,
        pop(pop_index,gene_index) = 0;
else
        pop(pop_index,gene_index) = 9;
end
     gene_index = gene_index+1;
% Subsequent Genes are significant digits temp(par_index) = abs(par
(pop_index,par_index))/10^dec(par_index);
for count = 1:sig(par_index),
            temp(par_index) = temp(par_index)*10;
            pop(pop_index,gene_index) = temp(par_index)-
                rem(temp(par_index),1);
            temp(par_index) = temp(par_index)-
                pop(pop_index,gene_index);
            gene_index = gene_index+1;
end
end
end
```

3.4 Selective

The basic principle of selection is that the more adaptive the chromosome, the greater the probability of selection. The selection not only determines which individuals are allowed to exist but also determines the number of possible offspring.

The best individual has a higher probability of selection than the less fit individual. This selection is so strong that the genes of the highly-adapted individual can prevail. In the opposite case, this selection also occurs, the genes of the low-adapted individual will be suppressed or low-dominance. This causes a local solution or premature convergence. If the selection is weaker, the less well-adapted individuals have a chance to reproduce. This will increase the probability of finding a global solution but slow down the evolution.

There are many selection methods such as proportional selection, round selection, linear ranking selection, and exponential ranking selection. In the content of this chapter, we use linear ranking selectivity. Linear rank selection is to arrange individuals in ascending order of fitness and assign the best individual class N, the worst individual class 1. The probability of selection of each individual is linearly proportional to its rank:

$$p_k = \frac{1}{N}\left[\eta + 2(1-\eta)\frac{k-1}{N-1}\right] \tag{7}$$

where $0 < \eta < 1$ the selection probabilities of the best and worst individuals are η/N and $(2-\eta)/N$. The program of the linear ranking selection process is as follows:

```
function parent = Select_Linear_Ranking(pop,fitness,eta,elitism,bestchrom)
% Linear selection
% input variables:
% pop: two-dimensional array [NxL], population before selection
% N is the population size, L is the length of the chromosomal chain
% fitness: column vector [Nx1], the fitness of individuals in a population
% elitism: The flag holds the superior individual in the process of evolution
% bestchrom: variable used to save the best individual in evolution
% eta: Parameters of linear rank selection
% Biên ra:
% Parent: two-dimensional array [NxL], population after selection
if (nargin<5), error('missing arguments');
end;
N = length(fitness); % population size
[fitness,order] = sort(fitness); % Arrange chromosomes in
  % ascending order of
 % fitness
% Calculate the probability of selection of chromosomes after sorting
according to formula (7)
for k = 1:N,
     p(k) = (eta+(2-eta)*(k-1)/(N-1))/N;
end
s = zeros(1,N + 1);
for k = 1:N,
     s(k + 1) = s(k) + p(k);
end;
for k = 1:N,
   if elitism ==1 & order(k) == bestchrom,
% If elitism = 1 and the chromosome in use is the best chromosome, then that
chromosome is selected
     parent(order(k),:) = pop(order(k),:);
else
% if elitism = 0 or the chromosome in use is not the best one
% then that chromosome will be selected with a probability
% proportional to its rank
          r = rand*s(N + 1);
          index = find(s < r);j = index(end);
          parent(order(k),:) = pop(order(j),:);
   end
end
```

3.5 Crossbreed

The more adaptive an individual is, the more likely it is to survive. Through the process of crossbreeding, the good traits of the previous generation are passed on to the next generation. Hybridization is a method of sharing information between chromosomes. This operation combines the characteristics of two parental chromosomes to produce two offspring with the prospect that a good parent will produce better offspring. Inbreeding usually does not affect all chromosomes but on the contrary, only occurs between two-parent chromosomes selected at random with a probability of hybridization. The principle of crossbreeding is to randomly pair two chromosomes in a population after having passed the selection step to create two daughter chromosomes, each child chromosome inherits a part of the parent's genes.

There are many different breeding methods such as single point crossbreeding, multiple point crossbreeding, and regular crossbreeding. Two-point crossbreeding means dividing each parent's chromosome into four random parts. When creating two son chromosomes, each son chromosome contains four parts including two parts of the father's chromosome and two parts of the mother's chromosome. The program of the breeding process is as follows:

```
function child = Cross_Twopoint(parent,Pc,elitism,bestchrom)
% Input Variables:
% parent: Two-dimensional array [NxL], population before
% hybridization
%N is the population size, L is chromosome length
%Pc: Probability of hybridization
% elitism: Flag that holds the superior individual in evolution
% bestchrom: The variable used to save the best individual in
% evolution
% Output Variables:
% child: two-dimensional array [NxL], population after
% hybridization
if(nargin <4),
    error([have not enough arguments']);
end;
[N,L] = size(parent);
for p1 = 1:N, % parent individual 1
% if elitism = 1 and p1 is a superior individual, do not cross
% (to preserve the best genes)
  if (elitism==1)&(p1==bestchrom)
      child(p1,:) = parent(p1,:);
else
if Pc > rand % hybridization occurs with
  % probability Pc
        p2 = p1; % Randomly select individual
  % parent 2 that is defferent
              % from individual parent 1
  while p2==p1,
              p2 = rand*N;
              p2 = p2-rem(p2,1) + 1;
end
            k1 = rand*(L-1); % random selection of
                        % hybrid point 1
            k1 = k1-rem(k1,1) + 1;
            k2 = k1;
while k2==k1,                % randomly select
```

```
                                        % hybrid point 2
                                        % defferent from
                                        % hybrid point 1
                        k2 = rand*(L-1);
                        k2 = k2-rem(k2,1) + 1;
end;
if k1 > k2, t = k2;k2 = k1;k1 = t; end;
% if k2 < k1 then convert k1 to k2 to ensure that k1 < k2
% Children inherit genes from their parents
            child(p1,1:k1) = parent(p1,1:k1);
            child(p1,k1 + 1:k2) = parent(p2,k1 + 1:k2);
            child(p1,k2 + 1:L) = parent(p1,k2 + 1:L);
else
            child(p1,:) = parent(p1,:);
end
end
end
```

3.6 Gene mutation

The offspring born through the process of natural selection and crossbreeding will carry the good qualities of the parent's generation. However, because the initial initialization generation is not rich and not suitable, the individuals cannot evenly spread the entire solution space. This makes it difficult to find the optimal solution. The mutation operation changes one or more genes of an individual to increase the diversity of the population structure. This helps to prevent the genetic algorithm from prematurely converging and the locally optimal solution. However, mutations should occur with low probability, otherwise, they can cause disturbance and loss of selected and highly adaptive individuals.

Mutation has many methods such as single point mutation, two-point mutation, and regular mutation. In this problem used uniform mutation. The program of the mutation process is as follows:

```
function newpop = Mutate_Uniform(pop,Pm,elitism,bestchrom)
% Input Variables:
% pop: Two-dimensional array [NxL], population before mutation
%N is population size, L is chromosome length
% Pm: Probability of mutation
% elitism: Flag used to preserve the superior individual in the process of
evolution
% bestchrom: The variable used to save the best instance in evolution
% Output variable:
% newpop: two-dimensional array [NxL], population after mutation
if(nargin<4),
    error(['missing arguments']);
end;
[N,L] = size(pop);
```

```
newpop = pop;
for pop_index = 1:N,
if(elitism==0)||(elitism==1&& pop_index~ = bestchrom),
for gene_index = 1:L,
  if Pm > rand,% The mutation occurs with the
              % probability that Pm generates
              % a random gene different from
              % the gene under consideration
        rand_gene = rand*10;
  while(pop(pop_index,gene_index)==
                rand_gene-rem(rand_gene,1)||rand_gene==10),
rand_gene = rand*10;
  end
              newpop(pop_index,gene_index)=
                rand_gene-rem(rand_gene,1);
  end
  end
  end
end
```

3.7 Convergence

If the preset number of generations is reached, stop breeding new generations and output an individual with the best chromosome sequence stored in the bestchrom memory cell. In this content, it is necessary to create five chromosomes, Kp1, Ki1, Kp2, Ki2, and Kd2. In case the previous generation objective function value is equal to the next generation objective function value in a preset number of generations, then also stop breeding and output the individual with the best chromosome sequence. The program of the convergence process is as follows:

```
% generation: variable that counts the number of generations
% terminal = 1: flag indicating the end of this algorithm
% stall_generation: variable that counts the number of generations of
adaptive functions
% check stop condition
if generation == max_generation
      Terminal = 1;
elseif generation >1,
  if abs(bestfit(generation)-bestfit(generation-1)) < epsilon,
stall_generation = stall_generation+1;
if stall_generation == max_stall_generation, Terminal =1;
end
else
          stall_generation =0; % variable count the
                               % number of generations
```

```
                                   % of fitness function
                                   % unchanged
    end;
    end;
  end;
```

4. The results of running the genetic algorithm to find the optimal solution

4.1 Parameters of the genetic algorithm

- The number of individuals in the initial population is 25.
- The maximum number of generations is 30.
- The number of generations with the same objective function result is 20.
- The number of chromosomes contained in each individual is 5 corresponding to two power controller parameters (G_{CN}) and using PI structure KP1, KI1, and three pressure controller parameters (G_{CP}) using PID structure are KP2, KI2, KD2.
- Initialization limit for parameters KP1, KI1, KP2, KI2, KD2 from 0 to 300.
- The objective function (J) includes the optimal parameters of the squared deviation of the signal of the setpoint transmit power and the actual transmitted power. Plus, the coal saving criterion is the squared ratio of the fuel flow to the actual generating capacity.

$$J = (\mathrm{Ne}^* - \mathrm{Ne})^2 + \left(\frac{w_f}{Ne}\right)^2 \rightarrow \min \tag{8}$$

- Probability of hybridization: 90%.
- Probability of mutation: 10%.

4.2 Results of the genetic algorithm after 30 generations of searching

The program that follows the algorithm of the genetic algorithm on MATLAB Simulink has been presented in Section 3. The value of the objective function in 30 generations is shown in **Figure 8**.

The value of the objective function reduce from 91.226 to 7911 after 30 generations, and the parameters of the two power and pressure controllers are found as follows:

- The G_{CN} power controller's parameter is K_{p1_out} = 4.3; K_{i1_out} = 0.017
- The G_{CP} pressure controller parameter is K_{p2_out} = 41; K_{i2_out} = 0.05; K_{d2_out} = 169

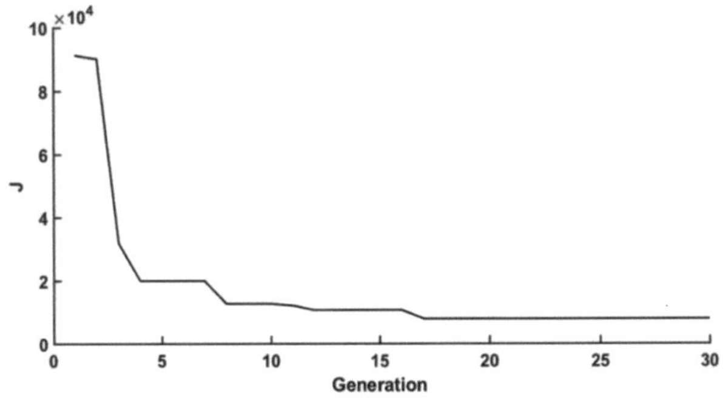

Figure 8.
Target J function decreases over 30 generations of the genetic algorithm.

While the old parameters of the controllers are as follows:

- The G_{CN} power controller's parameter is $K_{p1} = 0.093$; $K_{i1} = 0.0053$
- The G_{CP} pressure controller parameter is $K_{p2} = 1.18$; $K_{i2} = 0.085$; $K_{d2} = 98.1$

We use new parameters that are found by genetic algorithm for two controllers G_{CN} and G_{CP} into the thermal load control system model with a new coordinated control structure on MATLAB Simulink. Then evaluate and compare the model with these new parameters with the model when the controllers have not been optimized parameters by genetic algorithm. The results are made as in Section 5 of this chapter.

5. Simulation and evaluation

To demonstrate that the parameters of the steam pressure controllers G_{CP} and power controller G_{CN} are optimal. In this content, the chapter's authors study, test, and compare a new coordinated control structure that has parameters optimized by genetic algorithms with a new coordinated control structure using IMC_PID and PID self-tuner methods on MATLAB Simulink to find controller parameters. The simulation scenario is kept the same as when simulating the new coordinated control structure designed in Section 2 of this chapter and considering the case of the working system without interference. **Figure 9** is the simulation result of the combined control mode when using GA, including power response Ne (%), pressure response Ph (%), fuel flow response W_f (%), meet steam flow rate W_h (%), response to steam envelope water level H (m), respond to residual O_2 concentration (%), respond to combustion chamber pressure P_{bd} (Pa), respond to superheated steam temperature Th (°C).

5.1 Simulation results

The simulation scenario here is kept the same as when the simulation for the new coordinated control structure is designed in Section 2 of this chapter and considers the case of the working system without interference. **Figure 9** shows the simulation results of the combined control system when using GA, including power response Ne (%), pressure response P_h (%), fuel flow response W_f (%), response steam flow

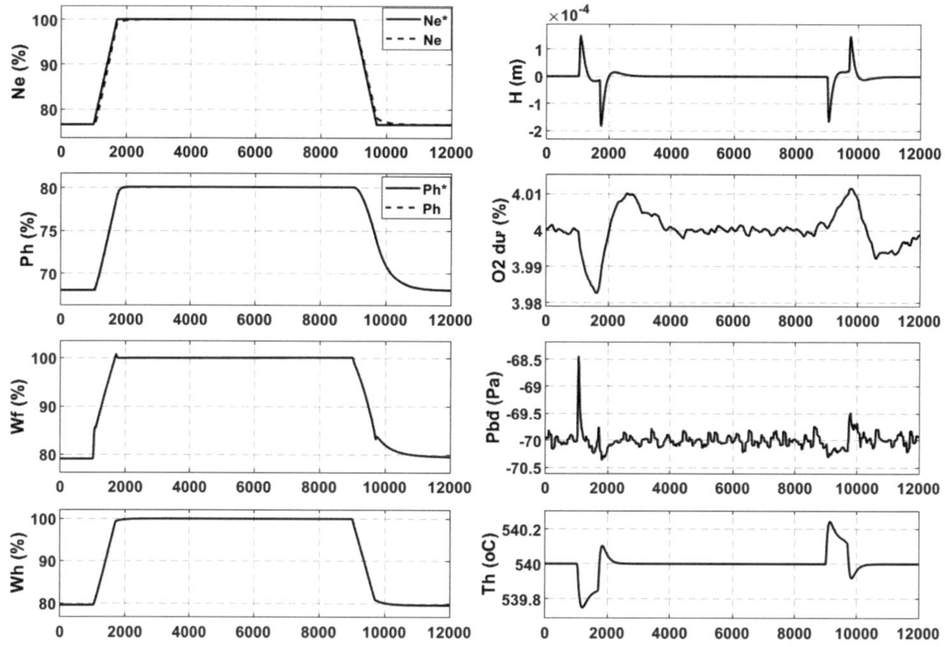

Figure 9.
Simulation responses of control mode using GA.

response W_h (%), response to steam envelope water level H (m), response to residual O_2 concentration (%), response to combustion chamber pressure Pbd (Pa), response to superheated steam temperature Th (°C).

We found that when reconfiguring the controllers according to GA, the responses of the four control loops in the boiler still ensure quality and stability.

We compare the response of the system using GA with the new coordinated control model designed by us and not using GA. Evaluation of the standard of transient time (Tqd), overshoot (%), maximum power deviation (eN max (%)), maximum pressure difference (eP max (%)) in load rise and fall time. **Table 2** gives a comparison of the power response:

Both cases are compared without overshoot. The coordinated control model using GA has better power response quality than this architecture without using GA. The results show that the response of the structure using GA in terms of transient time when increasing/decreasing the load and the maximum power difference is smaller (**Table 3**).

New coordinated control structure	T_{qd} (s)		e_{max} (%)	
	Increasing load	Decreasing load	Increasing load	Decreasing load
Not using GA	822	1155	8.06	4.94
Using GA	668	692	1.76	1.62

Table 2.
Evaluate the quality of power response.

New coordinated control structure	T_{qd} (s)		e_{max} (%)	
	Increasing load	Decreasing load	Increasing load	Decreasing load
Not using GA	812	2250	0.392	0.18
Using GA	660	2250	0.24	0.05

Table 3.
Evaluate the quality of pressure response.

Similar to the power response, the pressure response of the two cases to be compared has no overshoot. The combined control mode using GA has better pressure response quality in terms of transient times when the load is increased.

5.2 Evaluation with optimal operating standard

As mentioned above, this chapter only focuses on two objectives: fast-tracking the power setpoint (J_N) and saving fuel (J_f). Therefore, we compared these two criteria for the new coordinated control structure and this new coordinated control structure when using GA to optimize the controller parameters G_{CN} and G_{CP}, the results shown in **Table 4**:

New coordinated control structure	J_N (%)	J_f (kg/kWh)
Not using GA	2.83	0.4119
Using GA	2.63	0.41

Table 4.
Value of targets J_N (%), J_f (kg (coal)/kWh) from the simulation results.

From the results obtained in **Table 4**, it can be seen that the power tracking of the new coordinated control structure when GA is used is much smaller than that structure when GA is not used. This shows that the real power follows the setpoint when GA is used. The coal saving standard when using GA is smaller and saves 1.9 g/kWh. Thus, it is estimated that with a thermal power plant generating a capacity of 300 MW running 6000 hours/year with an output of 1.8 billion kWh, the coal saving amount of 1.9 g/kWh could have great economic value. The obtained results show that these parameters have been optimized by using the genetic algorithm GA.

6. Conclusions

The chapter has shown the advantages of using genetic algorithms in finding the optimal parameters of the controller in the thermal load control system in a coal-fired power plant. In this chapter, a genetic algorithm is designed for a new coordinated control structure with the goal of fast-tracking the set-point and fuel saving. Simulation modeling for these structures has been performed on MATLAB Simulink software. The structures evaluated which is the new coordination control structure proposed by the authors, the other optimized by using GA. The results show that all the three structures have good control quality, stability, and the new coordination control structure optimized by a genetic algorithm is much better than the other two structures compared to the targets of tracking the set-point quickly and saving fuel (the coal saving amount of 1.9 g/kWh).

Author details

PhamThi Ly[1*] and Bui Quoc Khanh[2*]

1 University of Transport and Communications, Hanoi, Vietnam

2 Institute for Control Engineering and Automation, Hanoi University of Science and Technology, Hanoi, Vietnam

*Address all correspondence to: ptlydk@utc.edu.vn and bqkhanh29@gmail.com

References

[1] Ventura S, Luna JM. Pattern Mining with Evolutionary Algorithms. Switzerland: Springer International Publishing; 2016. DOI: 10.1007/978-3-319-33858-3

[2] Sivanandam SN, Deepa SN. Introduction to Genetic Algorithms. Berlin, Heidelberg: Springer-Verlag; 2008. ISBN 978-3-540-73189-4

[3] Voigt H-M, Ebeling W, Rechenberg I, Schwefel H-P. New Genetic Local Search Operator for the Traveling Salesman Problem. University of Siegen; 1996. DOI: 10.1007/3-540-61723-X_1052

[4] Michalewicz Z. Genetic Algorithms + Data Structures = Evolution Programs. Berlin, Heidelberg: Springer-Verlag; 1994. DOI: 10.1007/978-3-662-07418-3

[5] Goldberg DE. Genetic Algorithms in Search, Optimization, and Machine Learning. Addison-Wesley Professional; 1989

[6] Garduno-Ramirez R, Lee KY. Multiobjective optimal power plant operation scheduling. IEEE Transactions on Energy Conversion. 2001;**16**(2): 115-122

[7] Technical Documents of the Hai Phong Thermal Power Plant. Haiphong Thermal Power Plant; 2017

[8] Technical Documents of the Expanded Uong Bi Thermal Power Plant. Uong Bi Thermal Power Plant; 2001

[9] Technical Documents of the Duyen Hai Thermal Power Plant. Duyen Hai Thermal Power Plant; 2018

[10] Jayachitra A, Vinodha R. Genetic algorithm based PID controller tuning approach for continuous stirred tank reactor. Advances in Artificial Intelligence. 2014;**2014**:1-8. DOI: 10.1155/2014/791230

[11] Malhotra R, Singh N, Singh Y. Genetic algorithms: Concepts, design for optimization of process controllers. Journal of Computing and Information Science in Engineering. 2011;**4**(2):39-54. DOI: 10.5539/cis.v4n2p39

[12] Saragih R. Parameters optimization of flexible structure using genetic algorithm. In: IEEE Int. Conf. Control Autom. ICCA; 2011. pp. 790-793. DOI: 10.1109/ICCA.2011.6137915

[13] Sanchez-Lopez A, Arroyo-Figueroa A, Villavicencio-Ramirez G. Intelligent control algorithm for steam temperature regulation of thermal power plants. In: Mex. Int. Conf. Artif. Intell. Berlin, Heidelberg: Springer; 2004. pp. 754-763

[14] Garduno-Ramirez R, Lee KY. A multiobjective-optimal neuro-fuzzy extension to power plant co-ordinated control. Transactions of the Institute of Measurement and Control. 2002; **24**(2):129-152. DOI: 10.1191/ 0142331202tm056oa

[15] Dhamanda A, Singh Rawat G. GA technique to solve the load frequency and tie-line power problem of thermal generating unit. Advanced Networks. 2019;7(2):51-58. DOI: 10.11648/ j.net.20190702.16

[16] Liu X, Guan P, Chan CW. Nonlinear multivariable power plant coordinate control by the constrained predictive scheme. IEEE Transactions on Control Systems Technology. 2010;**18**(5): 1116-1125. DOI: 10.1109/TCST.2009. 2034640

[17] Wang D, Huang B, Meng L, Han P. Predictive control for the boiler-turbine unit using ANFIS. In: Proc. Int. Symp. Test Meas. Vol. 2. 2009. pp. 1-4. DOI: 10.1109/ICTM.2009.5413102

[18] Flynn D, editor. Thermal Power Plant Simulation and Control. Stevenage, UK: IEE Press; 2003

Section 3

Other Applications

Chapter 7

Towards a Precise and Mathematical Fractalesque Architecture

John Charles Driscoll

Abstract

This paper reviews a design process in the context of algorithmic architecture design for establishing a scale-invariant and rigorous self-similar motif(s) that can be applied generally to any design problem. An architect (author) defines a genetic algorithm (GA) using a population of design variants iterated over multiple generations. Exemplars are selected based on their fractal dimension (FD) along with the architect and fit to solve a real-world architectural problem. The algorithm is coded in Python and Ruby with an interface in SketchUp. The architect is able to modify exemplars and iterate them as many times as required in the GA until an acceptable solution is achieved. Solutions are critiqued by a jury of professional architects regarding their fractal qualities. Results show a fractal motif that is not strictly self-similar and not strictly scale-invariant. Discussion is focused here on the philosophical implications of this research in terms of better defining a fractalesque architecture. The case for a more precise and mathematical fractalesque architecture is discussed concluding that further development of the algorithmic design process is necessary to clarify the value of such a tool.

Keywords: architecture, Genetic algorithm, algorithmic design, fractals, fractal dimension, box-counting dimension, characteristic complexity, motif, jury, critique

1. Introduction

This paper reviews a design process in the context of algorithmic architecture design for establishing a scale-invariant and rigorous self-similar motif that can be applied generally to any design problem. An architect (author) defines a genetic algorithm (GA) using a population of design variants iterated over multiple generations. Exemplars are selected based on their fractal dimension (FD) along with the architect and fit to solve a real-world architectural problem. This study diverges from some precedent in that it positions FD analysis up front from within the creative milieu of an architect's process. This study explores the use of FD from within a computational framework aimed at establishing both a scale-invariant and self-similar pattern as the organizing principle for the building's form and parti relative to various features of the building. These features are primarily plan, section and elevation as well as the relationship to a contextual building's relevant elevation. The algorithm is coded in

Python and Ruby with an interface in SketchUp. The framework is directed towards the creation of fractaleque architecture by incorporating a mathematically rigorous and iterative evolutionary strategy from within the creative milieu of an architect's process. The philosophical implications of the framework are the focus of this paper. The algorithmic design process consists of 3 steps as follows:

1. A GA is used to iterate a 2-dimensional line composition for a select number of runs. The GA uses a binary representation for individual lines which can be traded with other strings through crossover and mutation and inherited by subsequent generations. From an initial population of individuals exemplars are chosen based on their FD and tournament selection as well as cloning and used for creating additional population pools. This process is repeated many times until the FD of exemplars approaches a pre-selected FD. Line compositions are extruded into 3-dimensional massing models stochastically.

2. This step is referred to as *fitting*. GA outputs are imported into SketchUp, a solid modeling environment, and are available for a designer's review and selection relative to the sample problem. The designer is able to modify the exemplars chosen and use them as seeds for additional GA runs (repeating step 1). The GA and designer thereby establish a self-similar geometry that is reflected in various features of the design. Through the use of this tool the designer may potentially create a building with a similar FD throughout its scale range in terms of its orthogonal projections. 3 scale levels are used for this case study but the design process is not limited to a specific number of levels and could reflect a greater range of scales as the designer prefers but not an infinite range of scale due to practical limitations.

3. A jury of professional architects engage in a critique of the design at 3 stages during the process. The stages are referred to as "pin-ups" and consist of presentation images and renderings of both the GA outputs in the form of a timeline (**Figure 1**) and the designer's modifications. Jurors critiques are based in praxis and personal design philosophy.

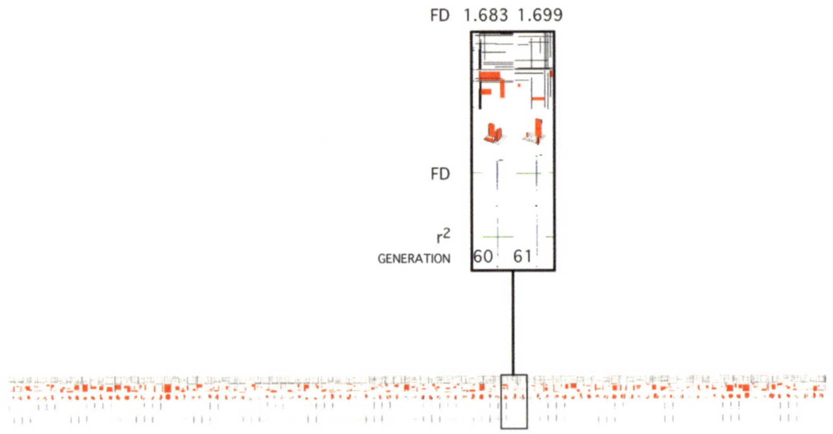

Figure 1.
Sample of timeline showing models progressing in order of iteration from left to right and increasing in BCD (image: Author).

2. Background

Fractal geometry was, from the beginning, closely related to natural inorganic and biological morphology [1]. Mandelbrot generalized a basis for what he termed "fractals" by gathering together the work of mathematicians and topologists such as Cantor, Gaston Julia, Felix Hausdorff, Jean Perrin and others who prior to computer technology had not been able to fully visualize the ramifications of such models [2]. Fractal dimension (FD) is a technique Mandelbrot adopted to describe an object's *characteristic complexity* and is related to its degree of irregularity at multiple scales although it does not explicitly define the object's geometry or whether or not it is strictly self-similar or fractal. The term characteristic complexity does however provide insight into an object's scale invariance. For this reason it is helpful in understanding a range of irregular geometries found in nature and architecture and is considered one of half a dozen or so measures of complexity [3].

Architecture has been trending towards ever larger buildings. From 2012 to 2018 the average floorspace of buildings has increased by11% in the United States according to the 2018 Energy Information Administration, Commercial Buildings Energy Consumption Survey [4]. Unfortunately, As buildings grow the morphology of buildings often tends to become simplified with less detail and less articulation (images of big box stores and strip malls come to mind). Larger buildings can be paradoxically cheaper through modularization and standardization as opposed to custom – one-off – buildings. With automated strategies on the horizon however it is increasingly possible to create a more articulate and detail rich architecture that is not prohibitively expensive.

FD incorporated in general analytic tools has been useful in understanding a building's detail in terms of characteristic complexity and have been gaining traction over the lasts decades. Michael Batty and Carl Bovil introduced FD as a serious tool in the analysis of architecture and urban form based on its underlying mathematical structure using box-counting dimension (BCD) to approximate FD (In this paper FD and BCD are used interchangeably). This tool is used to analyze 2-dimensional urban form and architectural plans and elevations of various buildings after they have been built [5, 6]. FD has since been adopted by many researchers to study the dynamics and complexity of cities and urban scaling as well as urban infrastructure [7–14]. Similar tools are also being used increasingly in the analysis of the characteristic complexity of architecture. FD is often used to make correlations between styles of architecture as well as the natural environment. Such analyses are related to fractal geometry a priori with respect to the traditional application of the tool, i.e., relating to morphogenesis. For this reason, architectural styles purporting to be based in nature are often a focus such as the American School or *organic* movement developed by Louis Sullivan and Frank Lloyd Wright [9, 15–23]. Otswald introduces a standardized calibrated model for the analysis of architecture in 2013 [24]. Lorenz *et al.* have offered additional analytic "proportion methods" using BCD offering a potentially more descriptive approach [25].

Algorithmic design or *computational design* (CD) uses computer technology to aid in generating architectural solutions and has developed rapidly over the last decades. CD approaches differ from simply using computer tools in the design process but rather use computation to create designs [26]. Generative design is a subset of CD and includes evolutionary approaches such as GAs [27–32]. GAs are methods analogous to biological evolution consisting of autonomous and stochastic search algorithms. GAs, when used generatively, can produce complex and unpredictable outcomes [26]. GAs are well suited to complex problems such as those represented by the multi-variate form and function requirements in architecture [33]. Trends in research are towards *biomimetic* approaches combining evolution-based computational methods with morphogenetic processes inspired by nature, where form is generated by computer technology, incorporating the rules and constraints of fabrication [34]. GA methods have been developed to help solve a variety of architectural problems such as geometric (lattice) deformation by Wattabe and sequences of polygonal operators by McGuire [35]. Latham and Todd developed the PC mutator system at IBM UK's Scientific Centre with individual projects as well as commercially available software [35]. Hemberg and the Emergent Design Group (EDG) at MIT developed Genr8 in 2001 which is a GA plug-in for the modeling and animation software Maya [36–38]. Galapagos is an evolutionary plug-in for Grasshopper [39, 40] which is used as a visual programming aid for architects within the BIM software Rhino. Grasshopper has been widely used for parametric modeling applications including a method for constructing islamic ornament [41] Galapagos has been used to optimize spatial adjacencies for complicated building programs [42], daylighting and shading studies [43] as well as to find novel solutions to structural problems [44]. Galapagos has also been used to generate new fractal forms for urban environments using cellular automata [45]. Chouchoulas *et al.* discuss Shape Grammars using GAs which have been explored with a prototype tool called Shape Evolution to design hypothetical buildings ([46], pp. 26–34). Shape grammars and parametric design has been applied to the architecture of Frank Lloyd Wright [33] Generative design, fractal geometry and stereotomic algorithms as well as the use of automated manufacturing processes have been influential to Joris Laarman and his designs for furniture and bridges [47]. Generative design incorporating FD has been developed as a method for designing efficient and robust structural forms [48–50]. Specifically related to this paper, generative design and FD has also been used to analyze and create neighborhood plans as well as individual schematic designs of residences [51].

3. Methodology

The case study was an architectural project in Ithaca, NY as defined by a request for proposals (RFP) issued by the city in the Spring of 2018 (https://www.cityofithaca.org/DocumentCenter/View/7614/Green-St-Garage-Redevelopment-IURA-RFP-Revised-5718). The project program consisted of a mixed use residential and retail building in an urban zone along Green Street to the south and with access to a pedestrian commons to the north. The case study focused on the schematic design phase with a level of completion considered to be adequate for an RFP. The design process consisted of a GA using BCD as the fitness criteria. The GA was written in Python and Ruby with a user interface plug-in for the solid modeling software SketchUp.

The GA started with a number of 2D line compositions representing *individuals*. Lines were placed horizontally and vertically on a page in random locations initially. Individual line compositions consisted of a set number of horizontal and vertical lines perpendicular or parallel to each other and the edge of the page. The lines are allowed to go from edge to edge or stop at another line. Each composition was measured with

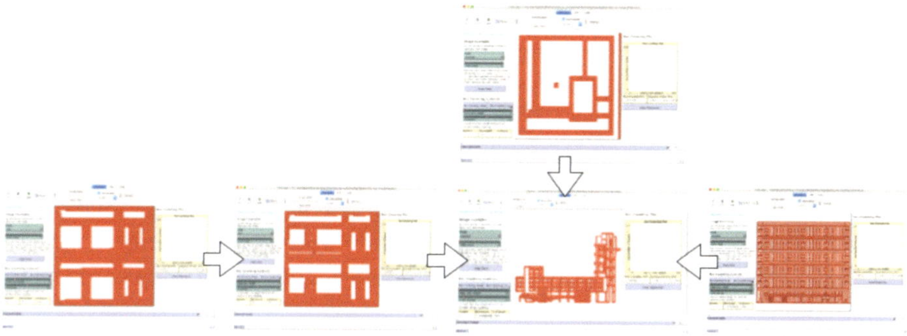

Figure 2.
BCD at 3 scale levels and adjacent building. Micro (above) = 1.267, mezzo (left) = 1.477, macro (middle) = 1.589, context (right) = 1.5616.

BCD to determine its scalar fitness value. The GA used *tournament selection* to search exemplar individuals whose value most closely approximated a target BCD. The target was 1.562 which was the BCD of the facade of the contextual building used in this study (**Figure 2**). Tournament selection consisted of randomly selecting 3 individuals from the pool with replacement and choosing the one whose value most closely matched the target. Tournaments were iterated until a new generation of individuals was created. New individuals were subject to random rates of mutation. Mutations consisted of altering the locations of lines an amount determined stochastically from a Poisson distribution. The full code for this research is available at https://github.com/JohnCDriscoll/Fractalesque-Architecture/new/main?readme=1.

Figure 1 shows a sample of a timeline which is out-put by the GA with the designs progressing from earlier designs on the left to later designs on the right. Exemplar compositions for each generation are shown along the upper-most row for each column. The images below the exemplar show the same line composition but with a number of rectangles defined by the lines stochastically selected as "masses"(shown in red). Masses are allowed to overlap and any number of masses may be set in the corresponding parameter. The row of images below show the masses extruded stochastically to an elevation along the Z axis. The graphs at the bottom display the exemplar FD for each run and the mean over successive generations and r-squared of individual designs. The FD is a measure of only the 2D compositions. **Figure 1** is an arbitrary example culled from preliminary GA runs. An important leverage point offered by a GA is the sheer number of potential variants a designer may peruse and develop. For this example the FD for exemplars highlighted are 1.683 and 1.699. As the series progresses to the right, the BCD increases. The GA presented here is understood as a proof-of-concept model and as such was explored briefly in terms of its capabilities. For instance, the runs executed were limited to 10 which was adequate to find a successful composition. Longer runs were executed with a target FD of 2 (maximum FD) as an experiment. A higher FD was achieved using 500 runs but the lines tended to bunch up along the edges of the page. This issue was not resolved in this research. Developing the GA further would focus on this issue as well as include more complex architectural elements other than simple 2D line compositions. This potential is discussed more in the Next Steps section below. As a proof-of-concept study these limitations were not considered significant. The GA developed in this research was intended to scale up in future versions and is presented here to demonstrate that an under-the-hood approach was valuable to the architect for a variety of reasons. A GA mirrors some aspects of the design process such as iteration and evolution and thereby allows for a more integrated approach where the architect is in control of the algorithm to a degree and enabled to visualize options from inception through realization. This is especially valuable in terms of fractal geometry which shares some of the attributes of the algorithm.

After the GA determined an exemplar composition it was out-pouted to a solid modeling environment where it was used to generate the design for the building. The designer made alterations and these designs were flatted to 2D compositions and used as seeds for the GA a second time. This back and forth happened 2 times in total. After this the designer modeled the building without the use of the GA. The completed design was measured for FD. Milestones during the process were presented to jurors as traditional architectural renderings. These milestones were referred to as *pin-ups*. There were 3 *pin-ups* in total. Comments were collected asynchronously and salient issues addressed and, in some cases, used to adjust the GA and fitting steps for subsequent GA and designer iterations leading to the final scheme.

4. Results

Results established a preliminary composition of 30 lines with a FD of 1.477. This result was somewhat lower than the target 1.562 but was significantly higher than random compositions which were in the 1.200 range. The composition was acceptable in terms of its design elements and potential constructibility. The lower FD was not considered problematic as a small scale element as will be discussed next. **Figure 3** shows a 2D image which was then extruded by the GA and modified by the architect. (**Figure 4**). This block became a module conceived as a masonry unit as well as the motif and was subsequently assigned scale-dependent functions and also established a general organizational strategy. The motif occurred at primarily 3 distinct scale ranges in the building termed *micro*, *mezzo* and *macro*. The micro level consisted of the modular element (**Figure 4**). The mezzo represented larger building systems such as facade elements and fenestration assemblies (**Figure 5**). The macro level was at the scale of the building overall as represented in elevation and plan including the site plan (**Figures 6–9**) The BCD of the various elements and scale ranges were coordinated within the architectural context of the site by approximating a similar BCD as the primary adjacent building at the macro level (**Figure 2**). The following description

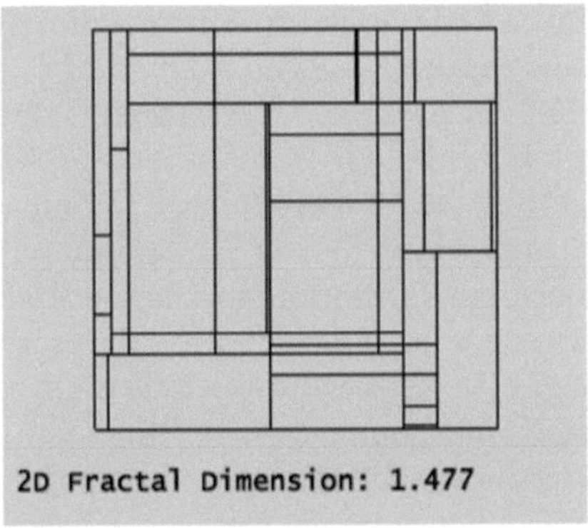

Figure 3.
Selected initial composition FD = 1.477 (image by author).

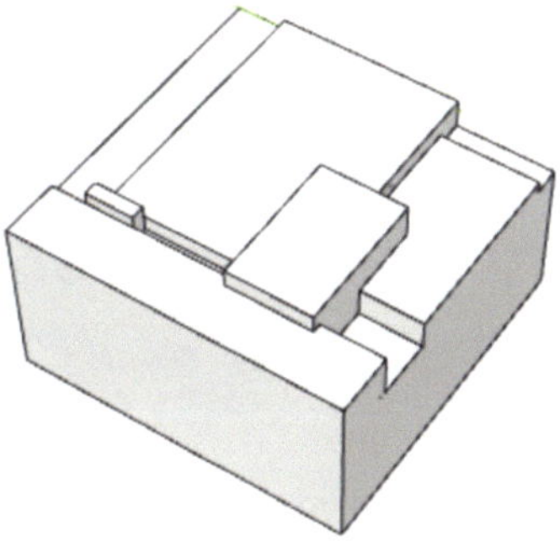

Figure 4.
Extruding and fitting of composition into a module used as a motif for the project. FD = 1.26 (image by author).

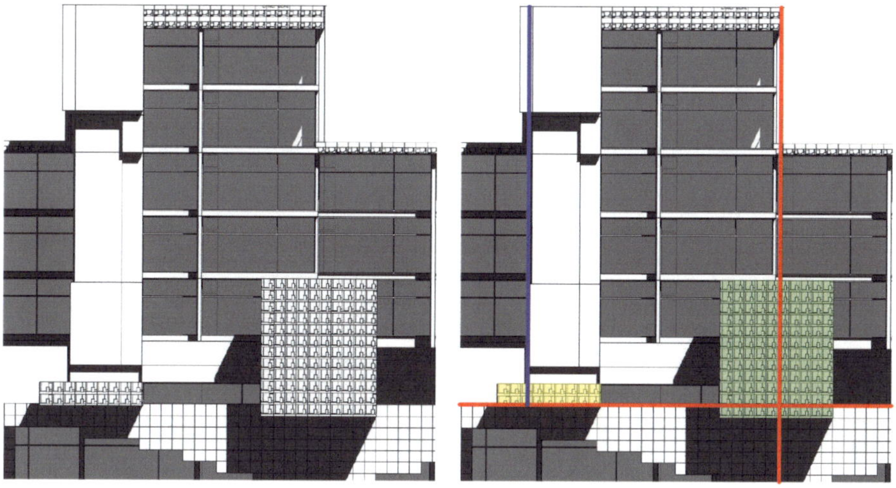

Figure 5.
Sample timeline of GA output using FD as fitness criterion. Image by author.

refers to **Figures 6–9**. Figures are shown as dual images with the left image representing the building element and the right image showing superimposed elements of composition which are discussed next.

The motif is composed of two intersecting lines (shown in red) with a third line (shown in blue) which stops at one of the intersecting lines. The main intersection of the red lines creates a visual focal point or center of interest. This point is then reinforced with a rectangle (shown in green) further establishing the focus. The intersection of the red and blue lines creates a secondary center of interest which is further established with the addition of the yellow rectangle (**Figure 10**). The micro level (**Figures 4** and **10**) represented a masonry unit and as such was simplified from the original GA output. The BCD reduced from 1.48 to 1.26. This was a significant reduction and not preferable. However, the requirements for forming the block outweighed the reduction in BCD in this instance.

The motif was again expressed at the mezzo level in various ways. A clear expression of the motif is shown in **Figure 5** where the proportions of the original block were preserved. The function of the element changes. Now, the primary focal point

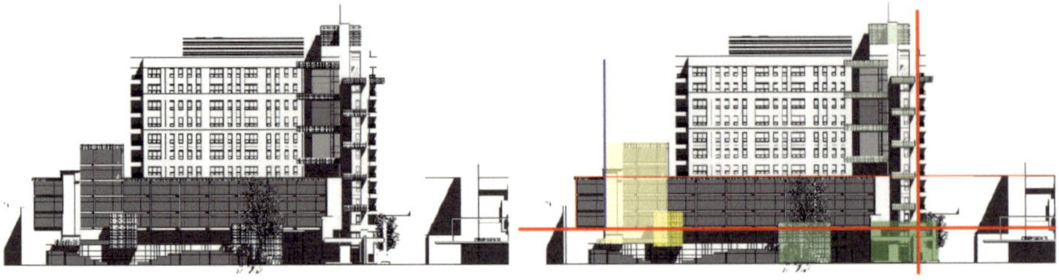

Figure 6.
Sample timeline of GA output using FD as fitness criterion. Image by author.

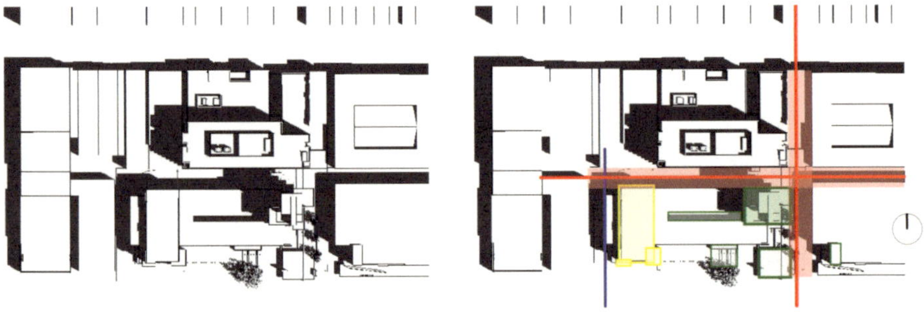

Figure 7.
Sample timeline of GA output using FD as fitness criterion. Image by author.

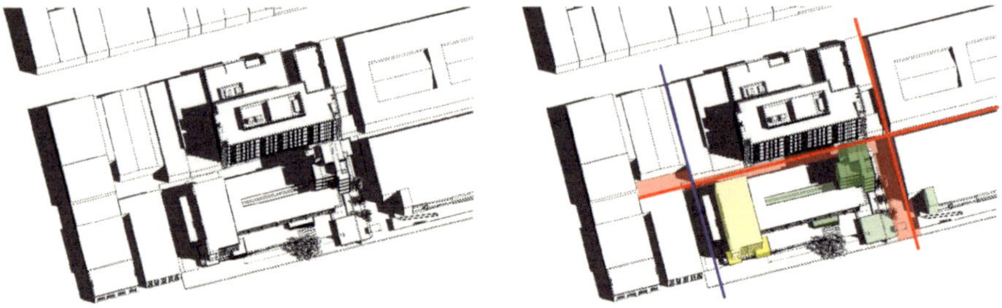

Figure 8.
Sample timeline of GA output using FD as fitness criterion. Image by author.

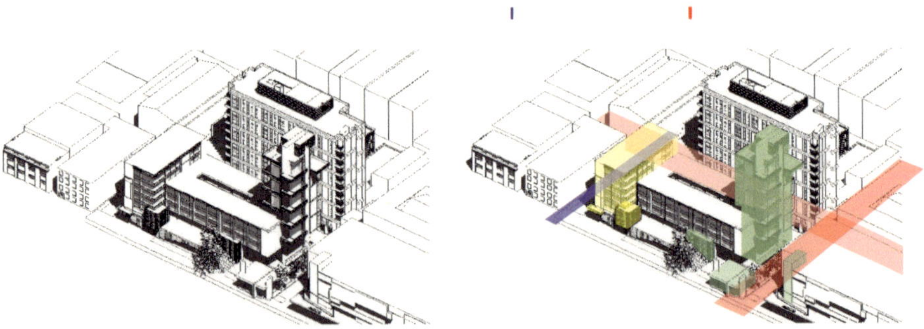

Figure 9.
Sample timeline of GA output using FD as fitness criterion. Image by author.

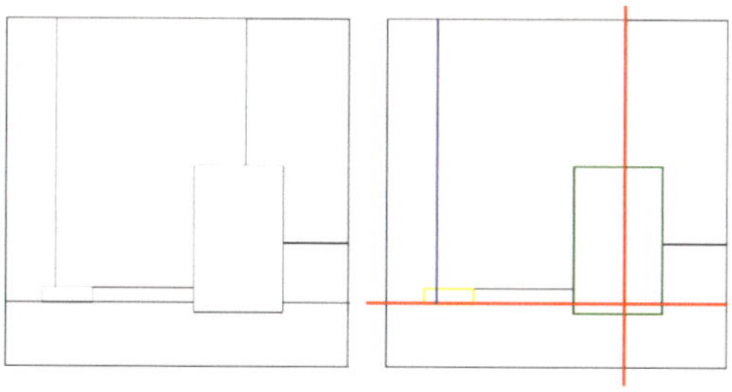

Figure 10.
Sample timeline of GA output using FD as fitness criterion. Image by author.

serves to denote one of the main entrances into the building leading to a primary circulation component while the secondary focal point serves to denote the entrance and storefront for a separate retail space. The red and blue lines outline the masses of the building and provide axes for the building to expand along. The mezzo scale was also included fenestration assemblies (**Figure 2**). Here the BCD was modified in a back and forth between architect and algorithm to increase the BCD back to the original GA output which was 1.48.

The macro scale was considered in terms of the entire south facade of the building as well as the site plan. The facade was measured and had a BCD of 1.59. This was higher than the original output but did compare favorably to the adjacent building which served as the contextual element for the building. This building had a south facade BCD of 1.56 which was considered a contextual relationship. Again, the elements of composition highlighted certain features of the building. These were mainly circulation routes and established a hierarchical arrangement of space which highlighted programmatic

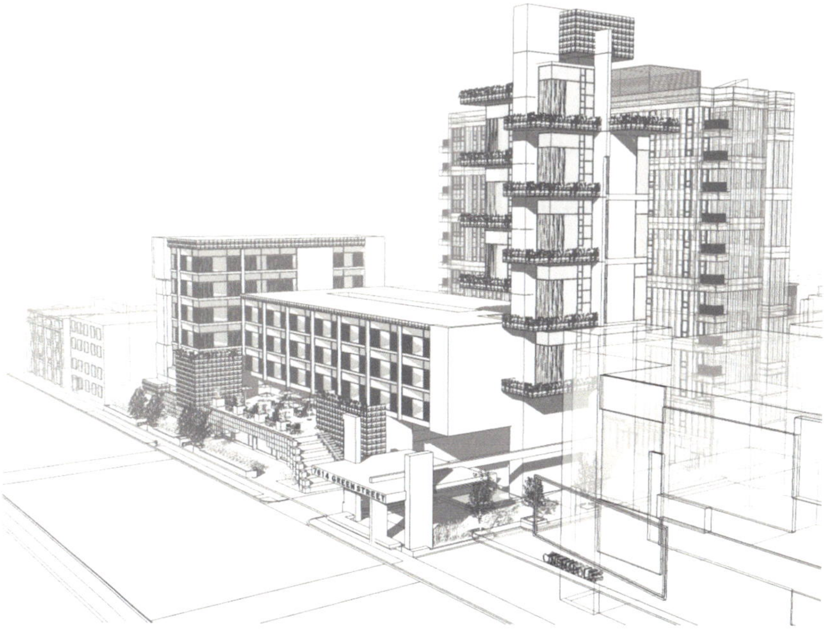

Figure 11.
Sample timeline of GA output using FD as fitness criterion. Image by author.

features of the building. The motif here serves to discern commercial from residential zones as well as demarcating the main access to the pedestrian commons located to the north. of the site (**Figure 6**). The site plan (**Figure 7**) shows this relationship more clearly where now the motif is superimposed on the ground plane. The access rout to the commons is well defined as well as the pedestrian strip between the building and the contextual building to the north. The green rectangles used to enhance the focal point are important aspects of the program, namely a park pavilion and the residential tower. The blue line is an important secondary feature of the building denoting a vehicular access point necessary for loading and delivery as well as egress and fire truck requirements. The next two images (**Figures 8** and **9**) are isometrics and show the same elements of composition and their 3-dimensional expression. **Figure 11** is a perspective rendering of the building and its relationship to the block. A key component of the building's program was to allow for views from the 11 story building to the north (contextual building) while also helping to balance the various proportions of the city block which consisted of smaller 4 and 5 story buildings. The inter-relations of scale included the building's components as well as the immediate urban fabric (**Figure 2**).

5. Discussion

As mentioned in the background section, previous studies have often discussed fractals in architecture within a context of an architectural tradition ascribing to nature based principles, e.g., the American School or *organic architecture* as espoused by Louis Sullivan and Frank Lloyd Wright [5, 15, 17, 18, 20, 52, 53]. This discussion continues in this tradition in focusing on the philosophical implications from a similar context of organic architecture with regard to the algorithmic method presented above.

James Walter Schildroth who was an apprentice of Frank Lloyd Wright's (see http://www.schildrotharchitect.net/autobiography.html) as well as a juror for this study writes in response to Harris' claims concerning the fractal ontology of Frank Lloyd Wright's design for the Palmer house [20]:

> *"I really don't see fractals in the Palmer House. Of course there are equilateral triangles. The unit is made with equilateral triangles. The plan is made by relating all parts of the plan as it is made to the unit system. The design is not made by repeatingself-similar triangles. The unit system gives the whole unity. I think this unity approaches the unity in all of the natural world. What we call beauty."*

A valid criticism from the jury was that the unifying motif is formal and did not originate from some functional requirement or relation to site but was established a priori and thereby applied functional requirements to the form, i.e., function followed form. This aspect of the process was built into the algorithm at the outset by limiting the fitness criterion to FD in selecting forms rather than a host of criteria such as: client's needs, budget, program, topography, solar attitude, materiality, tectonic systems, adjacencies, meaning, beauty, etc. This was done to simplify representation in the GA and generalize the system towards a parsimonious tool capable of solving a host of design problems. As the results show, the motif undergoes modification at different scales and as different functions are applied to it. The primary scale ranges employ the motif in different ways that are not exact replicas of each other although they are proportional as will be discussed. The building is therefore not self-similar in a strict sense. The scales range is limited to 3 levels although there is overlap and compositional unity throughout the building. Therefore, the building is clearly not scale-invariant, at least in terms of the repeating motif. The building may be considered as being composed of natural fractal structures in a material sense but not in so far as they were designed within the architectural concept. The building can not be said to be a literal representation in the mathematical or natural sense of fractal geometry in light of these two criteria not being met. The question of whether the building represents *fractal architecture* is more difficult to assess because of the ambiguity around this definition [53]. The following discussion does not argue that the building is fractal, fractalesque or fractal-like [54] but rather raises issues pertinent to this question.

One issue raised in this work and elsewhere in the literature is whether a higher FD is correlated with more quality design. Some suggest there is a "magic number" that is more esthetically pleasing or proportionally harmonious ranging from 1.3 to 1.52 [25]. The jury was mixed on this point with no clear indication that compositions with a specific FD were more compelling visually or had inherent merit architecturally when divorced from programmatic and functional requirements. Ken Kroeger comments, " To address the question about increased FD and improvement of the compositional quality – I don't think it specifically improves the composition. It simply is a variation of line work. Without having any performance/outcome values, I would bet that if you ask 50 different people which one they preferred, you would get that many answers."

This study however did not employ FD as an esthetic device or assume that one value of FD improved one composition over another but incorporated it from within an architect's design process to gauge and thereby relate the various elements within an overall organizational strategy. FD was considered on multiple levels from ornamentation on the facade (masonry blocks) to the distribution of spaces and circulation systems to the overall parti of the building and relationship to the larger urban fabric.

FD indicates an object's characteristic complexity but does not reveal the specific fractal shape. For instance, different fractals may have the same FD if they share the same characteristic complexity. This limitation however was not problematic in this study because the self-similar motif was defined by the GA and architect not by its FD alone. The GA can be thought of as a computational tool that takes over when the architect can no longer process. Discovering this edge and broadening it is a challenge for the architect. For this reason, FD used here does not present a conflagration of irregular objects with fractal objects. FD is not used to define self-similar geometries or indicate a higher quality design per se but is used compositionally to direct the eye.

A higher characteristic complexity in one element over another is useful in directing the eye [55]. In this study eye movement was an organizational strategy a priory and shown to be a convincing device in differentiating and highlighting the programatic elements as well as creating visual effect. The compositional elements of the motif established focal points and as such direct the eye towards key features of the building. The building as a whole is unified through the repetitive use of the motif and the dynamic quality established in the ensemble. The eye is lead from one place to another at various scale ranges in the building to discover the whole reflected in the part and vice versa. A juror, Bret Holverstott, comments, "I think I understand why our aesthetics seems to dictate that the fractal dimension increase as the scale increases. It is because the overall massing of a building should strive for a composition that is compelling if seen from a distance, not overly simple as in a homogenous skyscraper. I am reminded of how gothic cathedrals evolved to utilize alternating bays in order to keep your eye from immediately slipping to the end of the hall; good skyscrapers also break up the composition into something that allows your eye to linger on elements instead of slipping up to the top of the building."

In this study, the motif is seen to overlap or nest within itself as it changed scale. For instance, the primary focal point in **Figure 5** becomes the secondary focal point in **Figures 6** and 7. Such nesting characteristics define a proportional relationship between self-similar objects and are not, as Schildroth observes, simply repeating shapes at various scales. This type of expression is challenging to understand in terms of strict affine transformations, such as we see in geometric fractals like the Koch curve or Sierpinski triangle, but is germane to the overlapping richness in detail we see in architecture. Christopher Alexander discusses the multiplicity of readings and overlapping characteristics of architecture and urban planning in, A City is not a Tree [56] and demonstrates the ambiguity of mapping such characteristics. Therefore, a more refined definition of fractalesque architecture may deviate from a purely mathematical description in this regard as a description of natural fractals does, i.e., fractals in architecture are not strictly self-similar as they do not continue to iterate past a certain point and often translate in various ways or vary depending on their scale or compositional strategy, i.e., overlapping, layering, figure-ground reversal, etc. Self-similarity as explored here is emblematic of a *theme* which provides a sense of unity to the project yet importantly the theme undergoes *development* and *variation*. Although the motif is repeated in a finite way literally (3 levels) the self-similarity is extended phenomenologically in both directions, e.g., at a smaller scale in terms of the blocks materiality and at a larger scale in terns of the parti and master plan relating to the contextual urban fabric. In this way the building is a representation of fractal structure in two senses, both as a physical instantiation of a kind of limited self-similarity as well as a metaphysical metaphor representing a fractal structure extending beyond the building's spatial limits.

Perhaps in light of this discussion the use of FD in assessing and designing buildings is more provocative heuristically as a general tool that can be applied to many styles and design intentions. The juror Tom Mlynarski comments, "I think the most compelling thing about your work is this incorporation of fitness criteria rooted in FD. It seems FD can be used as a fitness criterion for all sorts of a different buildings with different styles and different programs and you can use the same criteria to rate them all. That is something fresh. You are no longer bound to rather simple fitness criteria like plan efficiency or exposure or whatever. So this is the most substantial part of your work." Mlynarski goes on to say, "... FD seems interesting in how it can be used to compare buildings that are very different in style and program. It's also interesting as a tool for judging options of the same design as you demonstrated here."

6. Next steps

As mentioned, an important aspect of this work was the parsimony of the GA/FD tool. Indeed, a GA was selected as an appropriate tool for its understandability and simplicity, i.e., it is autonomous and does not require training data. Similar to applying Occam's razor to the theory of evolution, GAs reflect a similar elegance and parsimony: essentially stochasticity within a constrained environment. However, GAs can become more complicated when multi-variate fitness functions are piled on. This study has endeavored to keep the fitness criteria simple – essentially one scalar metric – while still aiming to capture the complexity in architecture. For this reason it is not desirable to complicate FD analysis needlessly, however, an extension into a volumetric measure is a next step. 3-dimensional BCD tools exist in other fields and have been proposed for fractal analysis in architecture [25, 57–61]. Incorporating such a tool is necessary at this stage.

7. Conclusion

This paper reviewed an algorithmic design model in the context of architecture incorporating a GA using FD as its fitness criterion and discussed its broader implications relative to characteristic complexity and a more refined definition of fractalesque architecture. The GA is presented here to demonstrate that an under-the-hood approach was valuable to the architect for a variety of reasons. A GA mirrors some aspects of the design process such as iteration and evolution and thereby allows for a more integrated approach where the architect is in control of the algorithm to a degree. This approach enabled the architect to visualize options from inception within the algorithm through realization in model space. This is especially valuable in terms of fractal geometry which shares some of the attributes of a GA such as iteration and variation within a set of constraints. The design process developed in this research established a limited self-similar motif which was employed as an organizing principle to unify the building at multiple scales in the design. This research is novel in the sense that the organizing principles incorporate the algorithm as well as the traditional modes of design. The finished product expresses its making not only in terms of material, structure and craft but in terms of its code. The motif was not merely the pattern on the facade but included the organization of algorithmic content as well as space, structure and master plan.. The philosophical implications of this research suggest that a fractalesque architecture might be better conceptualized as both a partial instantiation of fractal geometry as well as in the metaphorical sense of a fractal which phenomenologically scales beyond the physical structure into its digital code and physical context. Such a design methodology required a facility on the part of the architect to incorporate the algorithm within a larger vocabulary. In essence internalizing it within the design process not unlike a material of sorts – in addition to brick and mortar – that the architect may sculpt to create a more mathematically rigorous self-similar motif with a consistent FD. The tool and process developed proved partially successful at approaching this benchmark. Further developing the GA/FD tool in the third dimension may improve the results.

Author details

John Charles Driscoll
Portland State University, Oregon, USA

*Address all correspondence to: driscoll.john92@gmail.com

References

[1] Mandelbrot B. How long is the coast of Britain? Statistical self-similarity and fractional dimension. Science. 1967;**156**(3775):636-638

[2] Mandelbrot BB, Mandelbrot BB. The Fractal Geometry of Nature. Vol. 1. New York: WH Freeman; 1982

[3] Mitchell M. Complexity: A Guided Tour. Oxford University Press; 2009

[4] IEA. Buildings energy consumption survey (CBECS). 2018. Available from: https://www.eia.gov/consumption/commercial/

[5] Bovill C. Fractal Geometry in Architecture and Design. Boston: Birkhäuser; 1996

[6] Batty M et al. Fractal Cities: A Geometry of Form and Function. Academic Press; 1994

[7] Driscoll JC. Fractals as Basis for Design and Critique. Portland State University; 2019

[8] Abundo C, Bodnar T, Driscoll J, Hatton I, Wright J. City population dynamics and fractal transport networks. In: Proceedings of the Santa Fe Institute's CSSS2013. 2013

[9] Abdelsalam M, Ibrahim M. Fractal dimension of islamic architecture: The case of the Mameluke Madrasas-Al-Sultan Hassan Madrasa. Gazi University Journal of Science. 2019;**32**(1):27-37

[10] Batty M. Cities and Complexity: Understanding Cities with Cellular Automata, Agent-Based Models, and Fractals. The MIT Press; 2007

[11] Bettencourt LMA et al. Growth, innovation, scaling, and the pace of life in cities. Proceedings of the National Academy of Sciences. 2007;**104**(17):7301-7306

[12] Encarnação S et al. Fractal cartography of urban areas. Scientific Reports. 2012;**2**(1):1-5

[13] Bettencourt LMA. The origins of scaling in cities. Science. 2013;**340**(6139):1438-1441

[14] Skrimizea E. Scale: The Universal Laws of Growth, Innovation, Sustainability, and the Pace of Life in Organisms, Cities, Economies, and Companies. 2021. pp. 184-187

[15] Lorenz WE. Fractal geometry of architecture. In: Biomimetics--Materials, Structures and Processes. Berlin, Heidelberg; 2011. pp. 179-200

[16] Ostwald MJ. "Fractal Architecture": Late twentieth century connections between architecture and fractal geometry. Nexus Network Journal. 2001;**3**(1):73-84

[17] Ostwald MJ, Vaughan J. The Fractal Dimension of Architecture. Vol. 1. Birkhäuser; 2016

[18] Vaughan J, Ostwald MJ. The relationship between the fractal dimension of plans and elevations in the architecture of Frank Lloyd Wright: Comparing the Prairie style, textile block and Usonian Periods. Architecture Science ArS. 2011;**4**(Dec):21-44

[19] Harris J. Integrated function systems and organic architecture from Wright to Mondrian. Nexus Network Journal. 2007;**9**(1):93-102

[20] Harris J. Fractal Architecture: Organic Design Philosophy in Theory and Practice. UNM Press; 2012

[21] Joye Y. Fractal architecture could be good for you. Nexus Network Journal. 2007;**9**(2):311-320

[22] Joye Y. A review of the presence and use of fractal geometry in architectural design. Environment and Planning B: Planning and Design. 2011;**38**(5):814-828

[23] Vaughan J, Ostwald MJ. Fractal geometry in architecture. In: Sriraman B, editor. Handbook of the Mathematics of the Arts and Sciences. 2018

[24] Ostwald MJ. The fractal analysis of architecture: Calibrating the box-counting method using scaling coefficient and grid disposition variables. Environment and Planning B: Planning and Design. 2013;**40**(4):644-663

[25] Lorenz WE, Andres J, Franck G. Fractal aesthetics in architecture. Applied Mathematics & Information Sciences. 2017;**11**(4):971-981

[26] Caetano I, Santos L, Leitão A. Computational design in architecture: Defining parametric, generative, and algorithmic design. Frontiers of Architectural Research. 2020;**9**(2):287-300

[27] Caldas L. Generation of energy-efficient architecture solutions applying GENE_ARCH: An evolution-based generative design system. Advanced Engineering Informatics. 2008;**22**(1):59-70

[28] Holland JH. Adaptation in Natural and Artificial Systems. Ann Arbor: University of Michigan Press; 1975

[29] Holland JH. Genetic algorithms. Scientific American. 1992;**267**(1):66-73

[30] Mitchell M. An Introduction to Genetic Algorithms. MIT Press; 1998

[31] Coates P, Broughton T, Jackson H. Exploring three-dimensional design worlds using lindenmayer systems and genetic programming. Evolutionary Design by Computers. 1999:323-341

[32] Coates P, Makris D. Genetic programming and spatial morphogenesis. AISB Symposium on Creative Evolutionary Systems, Edinburgh College of Art and Division of Informatics (AISB'99), University of Edinburgh, March 1999. 1999

[33] Granadeiro V, Pina L, Duarte JP, Correia JR, Leal VM. A general indirect representation for optimization of generative design systems by genetic algorithms: Application to a shape grammar-based design system. Automation in Construction. 2013;**35**:374-382

[34] Menges A. Biomimetic design processes in architecture: Morphogenetic and evolutionary computational design. Bioinspiration & Biomimetics. 2012;7(1):015003

[35] Romero JJ. The Art of Artificial Evolution: A Handbook on Evolutionary Art and Music. Springer Science & Business Media; 2008

[36] Hemberg M et al. Exploring generative growth and evolutionary computation for architectural design. In: Art of Artificial Evolution, Heidelberg. 2006

[37] Hemberg M et al. Genr8: Architects' experience with an emergent design tool. In: The Art of Artificial Evolution. Heidelberg. 2008. pp. 167-188

[38] Hemberg M, O'Reilly U-M. Extending grammatical evolution to evolve digital surfaces with genr8. In: European Conference on Genetic Programming. Berlin, Heidelberg: Springer; 2004

[39] Rutten D. Evolutionary Principles applied to Problem solving using Galapagos. In: AAG10, Vienna. 2010

[40] Rutten D. Galapagos: On the logic and limitations of generic solvers. Architectural Design. 2013;**83**(2):132-135

[41] Nadyrshine N, Nadyrshine L, Khafizov R, Ibragimova N, Mkhitarian K. Parametric methods for constructing the Islamic ornament. In: E3S Web of Conferences. Vol. 274. 2021

[42] Boon C et al. Optimizing spatial adjacencies using evolutionary parametric tools: Using Grasshopper and Galapagos to Analyze, Visualize, and Improve Complex Architectural Programming. Research Journal. 2015;7:25-37

[43] González J, Fiorito F. Daylight design of office buildings: Optimisation of external solar shadings by using combined simulation methods. Buildings. 2015;**5**(2):560-580

[44] Danhaive RA, Mueller CT. Combining parametric modeling and interactive optimization for high performance and creative structural design. In: Proceedings of the International Association for Shell and Spatial Structures (IASS). 2015

[45] Devetaković M et al. Fractal parametric models of urban spaces. Tehnički vjesnik. 2015;**22**(6):1547-1552

[46] Chouchoulas O, Day A. Design exploration using a shape grammar with a genetic algorithm Open House International. 2007;**32**(2):26-34

[47] Doubrovski Z, Verlinden JC, Geraedts JMP. Optimal design for additive manufacturing: Opportunities and challenges. International Design Engineering Technical Conferences and Computers and Information in Engineering Conference. Vol. 54860. 2011

[48] Rian I Md, Asayama S. Computational design of a nature-inspired architectural structure using the concepts of self-similar and random fractals. Automation in Construction. 2016;**66**:43-58

[49] Rian I Md, Sassone M, Asayama S. From fractal geometry to architecture: Designing a grid-shell-like structure using the Takagi–Landsberg surface. Computer-Aided Design. 2018;**98**:40-53

[50] Kiani Z, Amiriparyan P. The structural and spatial analysing of fractal geometry in organizing of Iranian traditional architecture. Procedia-Social and Behavioral Sciences. 2016;**216**:766-777

[51] Gürbüz E, Çağdaş G, Alaçam S. A generative design model for Gaziantep's traditional pattern. In: Proceedings of the 28th Conference on Education of Computer Aided Architectural Design in Europe. 2010

[52] Ostwald MJ, Vaughan J, Tucker C. Characteristic visual complexity: Fractal dimensions in the architecture of Frank Lloyd Wright and Le Corbusier. In: Architecture and Mathematics from Antiquity to the Future. 2015

[53] Ostwald MJ. Fractal architecture: The philosophical implications of an iterative design process. Communication and Cognition. 2003;**36**:263-296

[54] Lorenz W. Combining Complexity and Harmony by the Box-Counting Method – A comparison between entrance façades of the Pantheon in Rome and Il Redentore by Palladio. 2013. DOI: 10.13140/2.1.3100.4487

[55] Lee JH, Ostwald MJ. Fractal dimension calculation and visual

attention simulation: Assessing the visual character of an Architectural Façade. Buildings. 2021;**11**(4):163

[56] Alexander C. A city is not a tree. 1965. In Architectural Forum (No. 04). 1964

[57] Feranie S, Fauzi U, Bijaksana S. 3D fractal dimension and flow properties in the pore structure of geological rocks. Fractals. 2011;**19**(03):291-297

[58] Jiménez J, López AM, Cruz J, Esteban FJ, Navas J, Villoslada P, et al. A Web platform for the interactive visualization and analysis of the 3D fractal dimension of MRI data. Journal of Biomedical Informatics. 2014;**51**:176-190

[59] Krohn S, Froeling M, Leemans A, Ostwald D, Villoslada P, Finke C, et al. Evaluation of the 3D fractal dimension as a marker of structural brain complexity in multiple-acquisition MRI. Human Brain Mapping. 2019;**40**(11):3299-3320

[60] de Miras JR, Navas J, Villoslada P, Esteban FJ. UJA-3DFD: A program to compute the 3D fractal dimension from MRI data. Computer Methods and Programs in Biomedicine. 2011;**104**(3):452-460

[61] Tang D, Marangoni AG. 3D fractal dimension of fat crystal networks. Chemical Physics Letters. 2006;**433**(1-3):248-252